DECISION SCIENCES

MAKING BETTER DECISIONS

J. Tod Fetherling

Part of the Omnimics Series

Blue Dot Publishing

7011 Ellendale Drive

Brentwood, TN. 37027

ISBN: 979-8-9928531-5-5

May 29, 2026

CONTENTS

For every leader who has ever made a bold call,

watched the consequences unfold,

and wondered, with genuine uncertainty,

whether it was wisdom or luck that had guided their hand.

The answer matters more than most people dare to ask!

"The unexamined life is not worth living."

Socrates, Apology (399 BC)

"A good decision is based on knowledge, and not on numbers."

Plato, Phaedrus

"The mind is not a vessel to be filled, but a fire to be kindled."

Plutarch, On Listening to Lectures

"It is not enough to have a good mind; the main thing is to use it well."

René Descartes, Discourse on the Method (1637)

"In any moment of decision, the best thing you can do is the right thing, the next best thing is the wrong thing, and the worst thing you can do is nothing."

Theodore Roosevelt

"A reliable way to make people believe in falsehoods is frequent repetition, because familiarity is not easily distinguished from truth."

Daniel Kahneman, Thinking, Fast and Slow (2011)

ABOUT THE AUTHOR

J. Tod Fetherling is a healthcare executive, strategist, and entrepreneur with more than thirty-fives years of experience at the intersection of technology, data, and organizational transformation. His career has spanned provider health systems, health information networks, population health analytics platforms, consumer products, and emerging technology companies serving the healthcare sector.

At The Health Network, Fetherling led a decision to stream the first live birth online, a data-informed, reputationally complex choice that became the most-watched live event in the organization's history and established a model for digital health engagement that was widely emulated in subsequent years. At Perception Health, he deployed machine learning to map disease clusters and help hospitals pre-position services in at-risk communities, an early application of the population health intelligence principles that became central to the Vairos platform described in this book.

The Jim and Susan narrative that runs through this book, and through its predecessor volumes, Omnimics and Vairos, is drawn from the cumulative experience of those engagements: the boardroom conversations, the failed initiatives, the unexpected successes, and the moments of genuine organizational learning that distinguished the organizations that built lasting advantage from those that merely managed quarter-to-quarter performance.

Fetherling's work continues at the frontier of healthcare Decision Sciences, with particular focus on the governance, ethics, and accountability dimensions of AI-augmented decision-making in clinical and operational settings. He is a frequent speaker on Decision Sciences, organizational transformation, and the future of healthcare analytics.

Also by J. Tod Fetherling

Omnimics: Knowing the Answer Before You Ask the Question

Vairos: Data Driven Strategic Planning

Decision Sciences: Making Better Decisions

FOREWORD: THE QUESTION THAT CHANGES EVERYTHING

This story begins with a chair scraping back from a conference table.

You've been in that room. Perhaps you were the one who pushed back from the table, the CEO who leaned into a long silence after a consultant finished a presentation, or the chief medical officer who stared at a dashboard for the third time that week and felt a slow, creeping suspicion that all this information was somehow not helping.

Maybe you were the analyst who built the model and waited for someone in the top-tier management to actually use it and during that wait, your frustration kept growing.

As the chair scrapes back, someone stands up and says: *"Alright. What are we going to do?"*

This exact moment, the gap between knowing and deciding, is the subject of this book.

I have spent more than thirty years working inside and alongside healthcare organizations, and the pattern I have observed is so consistent it borders on the uncanny. Organizations invest enormously in data. They recruit talented analysts. They deploy sophisticated technology. They produce dashboards that look good but are a functional disaster because of how complex they are to use.

But somehow, almost every time, the plan starts to fall apart between the idea and the execution.

The decisions get made, but these decisions aren't good enough and they're not deliberate. Not in ways that can be tracked, reviewed, or learned from. Not in ways that compound institutional wisdom over time rather than simply surviving each quarter on a combination of effort, instinct, and luck.

The discipline that's missing has a name and it's called Decision Sciences.

Three Books, One Journey

This book is the third in a series that began with Omnimics, the story of building the data foundation, and continued with Vairos, the story of converting raw data into actionable insight. If you have read those books, you know Jim and Susan. You know MegaHealth, the regional health system where they have been working together to transform not just the technology infrastructure but the organizational culture around data.

If this is your first encounter with Jim and Susan, do not worry. While this book is part of a series, it does stand on its own.. But know that what you are entering is not merely a business methodology text. It is the final movement of a symphony that has been building across two prior volumes, and this movement, like all final movements, is where the full weight of everything that came before arrives at its resolution.

Jim Johnson has the data. He has the analytics. He has, thanks to years of painstaking infrastructure work, the insights. What he does not yet have, and what Susan Myles is about to give him, is the discipline to convert those insights into better decisions, track those decisions over time, and build, from the accumulation of choices and consequences, something rarer and more valuable than information.

He is about to build wisdom.

Why Decision Sciences Now

We live in what historians of science will likely describe as the golden age of measurement. More data is generated every two days than was generated in all of human history prior to 2003. The processing power to analyze it is doubling at rates that would have seemed implausible to anyone alive before the twenty-first century.

Artificial intelligence can now diagnose certain cancers with greater accuracy than trained radiologists, predict patient deterioration hours before clinical signs become apparent, and optimize supply chains in real time across global networks of extraordinary complexity.

And yet.

Major strategic initiatives fail at rates that have remained stubbornly consistent for decades, between 60 and 70 percent of significant organizational transformations fall short of their objectives, according to research published by McKinsey, Kotter, and dozens of independent academic studies. Diagnostic errors affect approximately 12 million Americans annually.

Hospital system mergers, despite being justified by detailed financial analyses, produce the anticipated synergies in fewer than half of cases. Pharmaceutical companies, working with some of the most sophisticated data infrastructure in human history, produce approved drugs that are later withdrawn for safety reasons at rates that suggest systematic failures of decision quality at multiple stages of development.

The problem is not a shortage of information. The problem is that we have not yet built, at the level of individual cognition, team process, or organizational culture, an adequate science of what to do with information and that is exactly what this book is all about.

What Decision Sciences Is, and Is Not

Decision Sciences is not a software platform, though technology plays a critical enabling role. It is not a consulting methodology, though professional services firms have built significant practices around its principles. It is not a synonym for data science, though the two disciplines are deeply interconnected and mutually reinforcing.

Decision Sciences is a discipline. It's a systematic, evidence-based approach to understanding how decisions are made, improving how they should be made, tracking what was decided and why, measuring whether those decisions produced the intended outcomes, and feeding those learnings back into future decision quality.

It draws from at least five distinct intellectual traditions: decision analysis and operations research, behavioral economics and cognitive psychology, causal inference and econometrics, neuroscience, and organizational behavior.

So, if you want to practice it with genuine sophistication, it will require a fluency across all five. This doesn't mean you have to be an expert in each one. However, you will need to understand when concepts and tools from each one are relevant and when they are not.

The chapters in this book are designed to help you develop that understanding and the story of Jim and Susan is here to show you what these concepts look like in practice when an organization faces stakes much like the real world.

A Note on the Philosophical Tradition

When it comes to business writing, people usually assume that being deep or philosophical doesn't help you get things done. They think bosses need checklists, not old ideas from guys like Plato. I completely disagree, and that's why I wrote this book.

You see Plato described the nature of knowledge, when he drew the distinction between doxa (mere opinion) and episteme (genuine knowledge), he was describing a problem that every analytics team and every executive team face every day.

The dashboard shows us what happened. The question of what it means, what we should do about it, and how confident we should be in our interpretation, those are questions of episteme, not doxa and the failure to take them seriously is one of the most expensive intellectual mistakes in modern organizational life.

When Aristotle wrote about phronesis, practical wisdom, the capacity to deliberate well about the human good, he was describing exactly the leadership quality that Decision Sciences aims to cultivate and institutionalize.

Make no mistake, this isn't about cleverness, analytical skill, domain expertise, etc. It's about the kind of seasoned, reflective, empirically grounded judgment that can make good decisions under conditions of genuine uncertainty.

When the Stoics counseled attention to what is within our control and acceptance of what is not, they were articulating a principle that modern decision analysts call the distinction between decision quality and outcome quality, a distinction that turns out to be one of the most important and most consistently ignored in organizational life.

The philosophical tradition runs through this book not as ornamentation but as foundation. These thinkers were grappling with the same problems we face. Their thinking, refined over millennia of argument and application, is genuinely useful.

How to Read This Book

Each chapter follows a consistent architecture: a philosophical opening that situates the topic in the longest available intellectual tradition; a narrative section that follows Jim and Susan as they apply the chapter's concepts in real time at MegaHealth; a scientific section that grounds the concepts in contemporary research; and a practical section that translates the concepts into tools, frameworks, and questions that any leader can use immediately.

As far as reading this book is concerned, you can choose to pinpoint a chapter that's most relevant to you because each chapter is designed to stand alone. But, reading it from the start is something I'd recommend because the journey builds throughout the book and story is much more meaningful and full of insights that way.

What I ask of you, regardless of how you read, is the same thing Susan asks of Jim in Chapter One: honesty. Not about your past decisions, those are done, and learning from them is a separate project. But honesty about the process you currently use to make decisions, and the gaps between what that process produces and what it could produce.

That is a gap larger than most people think and realizing that is the real starting point.

J. Tod Fetherling

PROLOGUE:
A COIN, A ROOM, AND THIRTY YEARS OF QUESTIONS

It is a Tuesday morning in April. Outside the fourteenth-floor windows of MegaHealth's administrative building, the river catches the early light in a way that is brief, but beautiful nonetheless. . Inside, Jim Johnson is staring at six dashboards and feeling a nameless unease that has been building for eighteen months.

He has more information than he has ever had. Better analysts. Better tools. A data infrastructure that costs $47 million to build and produces genuine analytical value every day. He has, by any external measure, succeeded at the first two phases of a transformation that he committed to a decade ago: building the data, then building the insight.

Today, he's sitting here at seven-fifty on a Tuesday morning, looking at six dashboards and feeling, not for the first time, not even for the twentieth, that he is still making his most important decisions the way he has always made them. With his gut, his instinct, the loudest voice in the room, and whatever he read on the plane last month.

Susan Myles arrives at eight-fifteen.

She reaches into her jacket pocket and takes out a coin.

What follows, in the eighteen months documented in this book, is the story of what happens when an organization decides to take seriously the question that should precede every strategic initiative, every capital investment, every clinical protocol decision, every staffing choice, and every market move:

- Not "what should we do?" but "how do we know if we are deciding well?"

The answer is more demanding, more interesting, and more humanly important than Jim expected when he sat down at his desk that morning.

And all of it began with a coin!

A note on method: This book traces the arc of a single eighteen-month Decision Sciences implementation at a fictional regional health system. Jim and Susan are composite characters drawn from more than thirty-five years of my work inside healthcare organizations. The MegaHealth Decision Sciences program, its failures, its insights, its governance structures, and its outcomes, is a plausible and research-grounded account of what such a program looks like in practice, not a case study of any specific organization. The research and philosophy that inform the narrative are real. The names are not.

CHAPTER ONE:
THE ANATOMY OF A DECISION

What we actually do – and What we think we do – When we choose

Plato's Cave and the Conference Room

Twenty-four hundred years ago, in a dialogue called the Republic, Plato described prisoners chained in a cave, facing a wall. Behind them burned a fire; between the fire and the prisoners, objects were carried back and forth, casting shadows on the wall. The prisoners had never seen anything but these shadows.

They had names for the shadows. They had developed entire systems of thought to explain the shadows' movements and predict their patterns. In their experience, the shadows were reality.

Plato's allegory is among the most discussed passages in the history of philosophy, and it has been applied to nearly every domain of human inquiry. But I want to apply it here, in a specific and practical way, to the experience of organizational decision-making in the age of data.

Consider the modern executive staring at a dashboard. The dashboard is a wall. The numbers on it, the revenue figures, the patient satisfaction scores, the readmission rates, the market share percentages.., all of them are shadows. They are not reality itself.

You see, all these metrics are representations of reality, filtered through choices about what to measure, how to measure it, how to display it, and how to interpret it.

So, in essence, all the executive is doing is describing and predicting the shadows just like the cavemen.

This is not a criticism. Shadows are all we ever have!

The question Plato was really asking, the question that has echoed through Western philosophy for two and a half millennia, is: given that we are working with representations rather than reality, how do we reason about them?? How do we distinguish the shadows that map reliably onto reality from the shadows that mislead us? How do we make good decisions under the fundamental epistemic constraint that we never have direct access to the things we are deciding about?

Decision Sciences is, in one sense, the practical and empirical answer to Plato's question. It is the discipline of reasoning well under the conditions of irreducible uncertainty and representation that define all human decision-making. It is not the elimination of the cave. It is the science of navigating it with maximum intelligence and minimum self-deception.

Aristotle's Three Kinds of Knowledge

Aristotle, Plato's most famous student, drew a distinction that remains perhaps the most useful in all of practical philosophy. He distinguished between three kinds of knowledge:

1. Episteme - We can see this as scientific or theoretical knowledge, the kind that can be demonstrated through logical argument and empirical observation,

2. Techne - This can be seen as craft or technical skill; that's embedded in doing something well.

3. Phronesis - This can be considered the highest form of practical knowledge, the capacity to deliberate well about what is good and what should be done in the circumstances of a specific, complex, real-world situation.

Modern organizations have invested heavily in episteme. The analytics team produces it daily. They have invested somewhat less in techne, the operational skills of execution. And they have invested almost nothing in phronesis, the discipline of wise deliberation.

This is the gap that Decision Sciences fills. Not more data. Not more analysis. Phronesis, the capacity to reason well, under conditions of complexity and uncertainty, about what should actually be done. Why? Because…

> *"We deliberate not about ends but about means. For a doctor does not deliberate whether he shall heal, nor an orator whether he shall persuade... they assume the end and consider how and by what means it is to be attained."*

Aristotle, Nicomachean Ethics, Book III

Aristotle's observation, that deliberation is about means, not ends, is one of the most practically useful insights in the history of decision theory. Organizations spend enormous energy debating and refining their strategic ends: mission statements, vision documents, and strategic plans.

However, they spend far less time building the deliberative machinery that converts stated ends into well-reasoned, well-tracked, well-evaluated decisions about means. This is why so many strategic plans fail at the execution stage: not because the ends were wrong, but because the decision architecture for achieving them was inadequate.

Jim: A Physician Who Became an Executive

Alright so now that we have some philosophical fundamentals in place, let's take a trip down memory lane and see how Jim became an executive.

Jim grew up in a small town of Wooster, Ohio, the son of a general practitioner who made house calls until the mid-1980s, long after most physicians had abandoned the practice. His father's car was a fixture on the roads of Wayne County, an aging Buick that smelled of antiseptic and coffee and became, for the young Jim, the symbol of medicine as a fundamentally personal act: one human being, with knowledge and commitment, going to where another human being was suffering, and doing something about it.

He went to medical school at Ohio State with this image firmly in mind. He was going to be like his father. He was going to know his patients by name, know their families, know their histories. He was going to practice medicine the way it was supposed to be practiced, as a relationship, not a transaction.

Then he graduated, completed his residency in internal medicine, and joined a hospital-based practice. Within three years, the reality of modern healthcare had collided with his father's model in ways that left him reeling. He was seeing thirty-two patients a day.

He had twelve minutes per appointment. His patients' medical records were scattered across multiple incompatible systems that could not talk to each other. He was making, he realized with growing alarm, dozens of decisions every day that were inadequately informed, not because he was a bad physician, but because the information he needed to be a good physician was systematically unavailable to him, or arrived too late, or arrived in a form that made it difficult to use.

In his fourth year of practice, Jim made a mistake…

There was a patient, a sixty-three-year-old man named Gerald, who had been seeing Jim for two years for hypertension and type 2 diabetes, came in complaining of fatigue and mild shortness of breath. Jim examined him, noted that his blood pressure was a bit higher than usual so he adjusted his medication, said he should take some rest, and come back if his condition got worse.

Twelve hours later, Gerald was in the emergency room with a myocardial infarction, a heart attack that could have taken his life. He survived. But Jim spent the next three months revisiting that appointment in his head., He kept asking himself what he had missed, what he had failed to ask, what signal was present in the data of that twelve-minute encounter that his pattern-matching had not recognized.

He had missed, he concluded, not for lack of knowledge but for lack of the right information, presented at the right time, in the right form. The fatigue was in the record, Gerald had mentioned it at the previous appointment as well, and Jim had attributed it to his diabetes management.

The shortness of breath was new, but Jim had been in the twelfth minute of a twelve-minute appointment, with a waiting room full of other patients, and his cognitive resources were not equal to the diagnostic complexity of that moment.

That is one diagnosis gone wrong didn't make Jim a worse physician, but it did change him. It made him something different: an executive. He entered healthcare administration with a physician's understanding of clinical stakes and a reformer's conviction that the system could be built better.

Jim believed that he could optimize the healthcare systems only if the:

- Information was organized in a different way.

- Decision support was available at the time of clinical need.

- Organizations gave practitioners what they needed to make good decisions.

Twenty years later, Jim was CEO of MegaHealth. He had built the data infrastructure, through the work described in Omnimics. He had built the strategic capability, through the work described in Vairos. And now, sitting in his conference room on a Tuesday morning in April, he was staring at a wall of dashboards and feeling, with the clarity of someone who has never quite shaken a thirty-year-old habit of self-examination, that something was still missing.

Susan: The Scientist Who Discovered the Limits of Science

Now that you're familiar with who Jim is, let's take a look at Susan…

She didn't grow up wanting to work in healthcare. She grew up wanting to understand patterns. As a child in suburban Boston, she was the kind of student who could not pass a bookstore without stopping at the mathematics section. She was the one who asked her parents, at age ten, why the weather forecast was sometimes wrong even when the meteorologist seemed confident. One day, she stayed after class in eighth grade to ask her stats teacher whether the regression line they had drawn through the scatter plot was the right one or just one of many possible right ones.

The teacher, a former actuary named Mr. Hendricks who had come to teaching late in life and brought with him an unusual depth of practical experience, paused before answering. *"That,"* he said, *"is exactly the right question. And the answer is that it depends on what you are trying to do, and what you are willing to assume."*

That answer stayed with Susan through a mathematics degree at MIT, a doctoral program in applied statistics at Carnegie Mellon, and a career that moved from financial services to technology to healthcare. What she had learned, through a decade of building models that were technically impressive and practically disappointing, was that the question *"what does the data tell us?"* is almost always less interesting than the question *"what will someone decide based on what the data tells us, and should they?"*

Susan had worked in finances and she knew how quantitative risk models had created a false sense of certainty. She had seen these models leading to decision failure first hand. She had worked in technology and saw how A/B led to better metrics while undermining the consequences. And in healthcare? Well, Susan had seen how predictive models were producing insights that physicians acknowledged, but disregarded when put alongside clinical intuition. The reason?

It's not that intuition was superior. It's that no one had bridged the gap between the model's output and the physician's decision process.

When Jim met her in a coffee house (if you've read the first book, you know. If not, no worries), this encounter led initially to the discussion about Omnimics data infrastructure project. She had accepted Jim's offer with one condition: she wanted to continue being a part of the organization but operate outside the four walls of the organization to be free of corporate politics.

She would engage long enough to build not just the data layer but the decision layer. She had seen too many organizations stop at insight and wanted to see, at least once in her career, what happened when an organization went further, MegaHealth seemed like the right option.

The Tuesday Meeting: Where Decision Sciences Begins

The conference room on the fourteenth floor of MegaHealth's administrative building has floor-to-ceiling windows on two sides. On a clear day you can see the river from the east-facing windows. Jim has never noticed the river because he's always facing the screen.

This Tuesday morning, the screens show six dashboards, all at the same time: patient volumes by service line, financial performance against budget, quality metrics, market share estimates, employee engagement scores, and, because the analytics team has been working on it for three months, a real-time map of population health indicators across the service area.

Susan enters at eight-fifteen, coffee in hand, laptop bag over one shoulder. She has been doing this long enough that she does not look at the screens immediately. She looks at Jim.

Susan says *"Tell me what's bothering you."* Jim is quiet for a moment. The river catches the morning light behind him and he says *"We have more information than we've ever had. We have better analytics than we've ever had. The Vairos system is doing things I literally could not have imagined five years ago. And I still feel like I'm making decisions the same way I always have. With my gut, and the loudest voice in the room, and whatever I remember reading in the last six months. Nothing has actually changed in how I decide."*

Hearing this Susan replies, *"That is the most honest thing you have said to me in four years of working together."*

Jim looks up and says, *"It doesn't feel like progress."*

Susan responds *"It is progress. It is the specific kind of progress that comes before real change, the moment when the person who needs to change understands clearly what needs changing. Most executives never get here. They buy the platform and declare victory. You are not doing that. You are asking the harder question."*

Jim, *"Which is what?"*

"How do I know if we are making better decisions?" Susan pauses, setting down her coffee and speaks

What Is a Decision? Getting the Definition Right

Before Susan explains what, she proposes to build, she makes Jim agree on a definition. This might seem a bit excessive, but it's not. Definitions matter enormously in Decision Sciences, because sloppy language about decisions leads to sloppy thinking about them and that is what leads to failures, the kind that fill business case studies at IVY league schools.

You see, when it comes to everyday business language, the word, *"decision"* is used loosely to mean anything from a momentary preference (I decided to order the salmon) to a lifelong commitment (I decided to become a physician).

In Decision Sciences, the word carries a more specific meaning.

Here, a decision is a commitment of resources, time, money, attention, reputation, or some combination, to one path over all available alternatives, made at a specific moment, by a specific agent or group, under specific conditions of information and uncertainty.

The commitment is what distinguishes a decision from a preference, a wish, or an opinion. This definition has four important implications that include: Decisions can be identified, recorded, and traced, they have a who, a when, a what, and a why.

1. Every decision has an opportunity cost, committing to one path means not committing to others, and the value of the unchosen paths is always part of the decision's true cost.

2. Decisions are made under specific information conditions that can be documented, making retrospective evaluation possible, which is the foundation of organizational learning.

3. The quality of a decision can be assessed separately from the quality of its outcome, a crucial distinction that we will return to repeatedly throughout this book.

Susan has Jim write this definition in the moleskin. She is a firm believer in writing as a cognitive discipline because when you write something, you are forced to see meaning in it, as opposed to just saying it vaguely.

39,000 Decisions a Day: The Scale of the Challenge

After Jim has written the definition, Susan tells him something that stops him cold.

"The average professional makes approximately 39,000 decisions a day." says Susan.

Hearing this Jim responds, *"That is impossible."*

Susan knows Jim is not ready to believe this insight, so she goes on to explain that *"It is not only possible, it is well documented. The confusion comes from conflating what we usually mean by a decision, a conscious deliberative choice, with what cognitive scientists mean by a decision, which includes every moment of directed attention, every motor sequence, every social inference, every perceptual categorization. Most of these are not deliberate in any meaningful sense. Your nervous system makes thousands of micro-decisions every second just to keep you upright and oriented in space. But even filtering purely cognitive decisions, choices about where to direct attention, what to believe, what to communicate, what to do, the number is in the thousands per day for any professional in a demanding role."*

At this moment Jim is a bit skeptical, but he's still listening as Susan continues to say: , *"Let me give you a simple illustration. You drive to this building every morning. That drive involves hundreds of decisions, when to change lanes, when to brake, how much distance to maintain from the car ahead. You make none of them consciously. They are fully automated by years of practice. Your prefrontal cortex is not involved. Your basal ganglia are handling it. Now how many of your decisions as CEO are as fully automated as your lane changes?"*

Jim, after a long pause, replies *"Probably more than I'd like to admit."*

To this Susan says *"And fewer than you think. That is exactly the problem. The decisions that feel automatic often are not. They feel automatic because they are fast, because they are consistent with past patterns, and because they do not generate conscious discomfort. But they are not actually operating on well-*

calibrated heuristics. They are operating on habits, on authority gradients, on whatever information happened to be salient that morning. The research on this is extensive and, if you will pardon the word, damning." Susan points to a slide that says:

The Neuroscience of the 39,000

Cognitive scientists Roy Baumeister and John Tierney, in their research on what they call "decision fatigue," demonstrated that the quality of human decision-making declines predictably over the course of a decision-making session.

Baumeister's famous study of judicial parole decisions, spanning 1,112 decisions across 50 case-days, found that favorable rulings declined from approximately 65% at the start of a session to nearly 0% immediately before breaks, then reset after food.

The implication: decision quality is not a stable trait. It is a resource that depletes. For a CEO making dozens of significant decisions daily, the cognitive architecture of when decisions are made may be as important as how they are made.

The Coin Flip: A Masterclass in Humility

Susan reaches into her jacket pocket and produces a quarter. The same one we mentioned at the start of the book. She sets it on the table in front of Jim with a small, precise motion. Then she opens a moleskin notebook and slides it across the table.

"I want to start with an exercise. We are going to flip this coin ten times. Each time, before I flip it, you are going to predict heads or tails, and you are going to write your prediction in the notebook. Then we will write the outcome. At the end, we will analyze what happened." - says Susan.

Jim, slightly amused and confused at the same time, said, *"This is Decision Sciences?"*

Susan replied, *"This is how Decision Sciences begins. With the humbling recognition that our confidence in our predictions is almost always higher than the predictions' actual accuracy. Pick: heads or tails?"*

Jim picks tail as Susan flips the coin. It's heads. She waits for Jim to write and then says *Your first bad decision. Note the date and time as well. Decisions have timestamps."*

They repeat the exercise. Jim picks tails again. The coin lands heads.

Susan, *"That is interesting. Why tails again?"*

Jim, *"I thought... if I got it wrong twice in a row, it would be more likely to come up tails next time."*

Hearing this Susan explained, *"That is the gambler's fallacy. It is one of the most persistent cognitive errors in human decision-making, and it has a neurological basis: your brain is designed to detect patterns and predict regularities, and it will confabulate patterns even in genuinely random sequences. The coin does not remember what it did last time. But your brain insists on treating the sequence as meaningful."*

They complete all ten flips. Jim gets six right and four wrong. Jim, *"Six out of ten. That's not bad."*

Susan, in a very confident and corrective tone, says *"If we ran 100 trials, you would converge on 50 percent. That is a random chance. The six-of-ten result is not evidence of predictive skill. It is a small sample with results consistent with pure randomness. And here is the critical question: if you had been making real decisions, not coin flips but healthcare strategy decisions, with the same confidence you felt before each prediction, what would that mean for the quality of those decisions?"*

This insight hit Jim right where it mattered. After this, Susan moved on to the next slide that told the story of how Portland, Oregon was named. .

In one of the most famous coin flips, Portland, Oregon's name was determined by a coin toss.

Two New England natives, Asa Lovejoy and Francis Pettygrove had to make the decision as they both owned the area.

They had their hometowns in mind for potential names. Lovejoy was from Boston, Massachusetts, while Pettygrove was from Portland, Maine. Pettygrove won the coin toss two out of three times and the rest as they say is history.

The coin used for this decision, now known as the Portland Penny, is on display in the headquarters of the Oregon Historical Society.

The coin flip exercise is not trivial. It is designed to demonstrate, in the most stripped-down possible way, the gap between the confidence one feels and the calibrated accuracy.

This gap, which decision scientists call overconfidence, is among the most consequential and most consistent findings in all of behavioral decision research. Decades of evidence, from meteorologists to physicians to military commanders, show that stated confidence reliably exceeds actual accuracy.

The question is not whether you are overconfident. It is whether your decision process compensates for the fact that you are.

Rock, Paper, Scissors: Competition, Prediction, and the Cognitive Trap

Susan does not stop with the coin. She reaches into her bag and produces a second moleskin, this one ruled, with a simple table she has already drawn: three columns labeled My Choice, Susan's Choice, and Winner.

Susan, *"We are going to play Rock, Paper, Scissors. Ten rounds. Same protocol, write your prediction, what you intend to throw, before each round. Then we play, and you record the outcome. I want you to track both your choices and mine."*

Jim, *"You are going to try to beat me, right? This is not like the coin."*

Susan, *"I am going to try to beat you. Yes. That is exactly right. This is different from the coin. The coin is a purely random environment. Rock, Paper, Scissors is a competitive environment with an adversarial agent whose behavior you are trying to predict. And this difference, which might seem small, turns out to be enormously revealing when it comes to how human decision-making works, and fails, in competitive contexts."*

They play. Susan wins six of the ten rounds, with two losses and two ties.

Round	Jim's Choice	Susan's Choice	Result
1	Rock	Paper	Susan Wins
2	Paper	Scissors	Susan Wins
3	Scissors	Scissors	Tie
4	Scissors	Paper	Jim Wins
5	Rock	Paper	Susan Wins
6	Paper	Scissors	Susan Wins
7	Scissors	Scissors	Tie
8	Scissors	Paper	Jim Wins
9	Paper	Rock	Jim Wins
10	Paper	Paper	Tie

Susan prompted Jim, *"Now I want to ask you a question. Why did you keep changing your choice in the first three rounds?"*

Jim responded *"I was trying to anticipate what you would pick."*

Susan said *"Based on what?"*

This question landed differently than the coin flip question. It carried more weight. Jim looked at the table in front of him. Rock. Paper. Scissors. Susan wins. Rock. Paper. Scissors. Susan wins again. He had been doing something, reading her, anticipating her, but now, asked to articulate the basis for it, he found that there was nothing solid there.

He had been reading her the way you read anyone when you are competing with them: watching her face, her hands, her posture, constructing a model of her intentions from signals that he now recognized were not actually signals. They were noise that his pattern-recognition machinery had transformed into a signal because that is what pattern-recognition machinery does.

, "I... genuinely don't know. I thought I was being strategic. But when you ask me to account for it, I can't." replied Jim.

Susan responded *"That answer just earned you a significant amount of my professional respect. Most people, when confronted with this question, immediately confabulate a reason. They say they read my body language, or they noticed a pattern in my first two throws, or some other post-hoc rationalization that sounds plausible but has no genuine predictive validity. You sat with uncertainty. That is unusual, and it is the cognitive posture that Decision Sciences requires."*

Susan explained the concept on a slide that said:

Hyperscanning and the Competitive Brain

A landmark study published in Frontiers in Human Neuroscience by Moerel, Varlet, and Grootswagers at Western Sydney University used hyperscanning, the simultaneous recording of brain activity from two interacting individuals, to study decision-making in competitive contexts.

Across 15,000 rounds of Rock, Paper, Scissors involving multiple pairs of participants, the researchers made three critical findings.

1. Players were systematically non-random: most overplayed *"Rock"*, the most instinctive, aggressive choice, followed by *"Paper,"* with *"Scissors"* least frequent.

2. Players consistently avoided repeating their previous choice, even though repetition would have been strategically equivalent to any other option in a random context.

3. The neural encoding of past outcomes predicted future choices, but only for losing players.

The brains of winners showed little encoding of outcome history. Losers were trapped in retrospective analysis. Winners had found a way to be more present. The organizational implication is direct: leaders who overanalyze past failures are systematically disadvantaged relative to those who can maintain forward-looking, adaptive thinking.

Explaining the slide, Susan said, *"Here is what the research shows. The optimal strategy in Rock, Paper, Scissors, if you want to win over many rounds against a human opponent, is to be genuinely random. But humans cannot be genuinely random. Our brains are built to detect and generate patterns, and when we try to be random, we produce predictable biases. We overplay Rock. We avoid repeating our last choice. We model our opponent's thinking and adjust based on recent outcomes, which sounds strategic but actually makes us more predictable, not less."*

Jim, *"So the competitive advantage goes to the person who can think less like a human being."*

Susan, *"In a limited sense, yes. The competitive advantage goes to the person whose decisions are harder to model and anticipate. In a narrow competitive context, that means being less patterned. But in a broader strategic context, healthcare market competition, for instance, it means something more interesting: it means being the organization that understands its own cognitive patterns well enough to compensate for them, while exploiting the predictable cognitive patterns of competitors."*

Jim, *"You can teach me to win?"*

Susan, *"I can teach you to make better decisions. Winning follows from that, but it is not guaranteed, because outcomes always contain elements beyond your control. This is one of the most important distinctions in all of Decision Sciences: decision quality and outcome quality are not the same thing and conflating them is one of the most expensive mistakes an organization can make."*

Decision Quality versus Outcome Quality: A Crucial Distinction

This distinction deserves its own extended discussion, because it is both counterintuitive and foundational to everything that follows.

In most workplaces, people think that if a result is good, the decision behind it must have been smart. If a plan brings in more money or wins more customers, we celebrate the choice as a "success." On the flip side, if things go south, we assume the decision was a mistake.

While it's human nature to think this way, it's actually a total misunderstanding of how things work. Judging a decision only by how it turned out is usually a mistake.

Consider the physician who decides not to order a cardiac stress test for a patient presenting with atypical symptoms. The patient has no heart attack.

The physician feels, and is perceived to be, vindicated. The decision was good, it avoided an unnecessary test and its costs and risks.

But was it? If the physician's decision-making process was flawed, if she failed to adequately consider the patient's risk factors, if she was operating on a biased heuristic, if she was in the fourteenth minute of a twelve-minute appointment, then the good outcome does not redeem the poor process. It merely obscures it. The next patient, processed through the same flawed decision-making architecture, may not be as lucky.

Now consider the opposite case: a hospital executive who makes an evidence-based, carefully analyzed decision to expand into a new market, only to have the expansion fail because a major employer unexpectedly relocated its operations out of the service area.

The outcome was poor. But if the decision process was sound, if it correctly assessed the available evidence, appropriately quantified the uncertainty, and chose the option with the best risk-adjusted expected value, then the poor outcome does not indict the process. It reveals the irreducible role of chance in any outcome-producing decision under uncertainty.

"A good decision is one made with the right process, the right information, and honest acknowledgment of uncertainty, regardless of what happens afterward. An organization that conflates good outcomes with good decisions will eventually suffer from a dangerous asymmetry: rewarding lucky thinking and punishing unlucky good thinking, until the good thinkers leave and only the lucky survivors remain.", Susan said to Jim, =

This way of thinking changes how companies learn from their mistakes and wins.

If we only look at what went wrong, we're only seeing half the story. This means that failures are often a mix of bad choices and just plain bad luck. If we only look at what went right, we're still missing the point, because those wins might just be a mix of good habits and lucky breaks.

To actually get better, we have to stop looking at the end result and start looking at how we made the choice in the first place. We need to study our decision-making process consistently over a long period of time, regardless of whether the outcome was a "win" or a "loss."

This is what Susan's proposed decision tracking infrastructure will make possible. Not just a record of what happened, but a record of what was decided, why, what was expected, and how reality compared to expectation. The feedback loop is not merely gratifying. It is the engine of organizational intelligence.

Building on What Came Before: Omnimics, Vairos, and the Third Layer

Jim sets down the moleskin and leans back in his chair. The morning has already produced more for him to think about than he had expected when he sat down with his first cup of coffee.

Jim, *"Omnimics taught me about the data. About where it comes from, how to trust it, how to govern it. Vairos taught me about analytics, how to turn data into something I could use. What are you going to teach me here?"*

Susan, *"I am going to teach you that data and analytics, as good as they are, are inputs to a decision process, and that the decision process is what actually determines outcomes. We have built the best data infrastructure in this region. We have built world-class analytics capabilities. And we have been stopping at the edge of the decision, watching insights evaporate into committee discussions and gut calls and org-chart negotiations. I want to build the layer that comes after insight."*

Jim, *"And what does that layer do?"*

Susan, *"It tracks decisions. It records them, who made them, on what evidence, with what alternatives considered, under what assumptions, with what expected outcomes. It measures those decisions against their actual results. And it feeds that measurement back into future decision-making. It is CQI, continuous quality improvement, applied not just to clinical processes but to the judgment that creates and evaluates those processes."*

Jim, *"That sounds like it could be uncomfortable."*

Susan, *"Extremely uncomfortable. Particularly for the people at the top of the organization, who currently make the most consequential decisions with the least systematic accountability. Including you. Especially you."*

Jim smiles. It is the smile of someone who has spent enough time with Susan to know that she is right, and that being right is her most consistent characteristic.

Jim, *"OK. Where do we start?"*

Susan, *"We start with you. I am going to bring a framework next week. But today, I want to do one more exercise. Are you familiar with a real decision you need to make in the next thirty days?"*

Jim, *"Yes. I need to decide about the radiation therapy equipment. A five-million-dollar decision, minimum. Possibly fifty million if we do the full fleet."*

Susan, *"Perfect. We'll use that. But not today. Today, we are still building the muscle. The moleskin goes with you. Every significant decision you make this week, write it down. What was the decision, what were the alternatives, what did you choose, why, and what are you expecting to happen. We'll review it next week."*

Jim picks up the moleskin. He turns it over in his hands. It is a small, practical object, the kind of thing he would have dismissed as a management-consulting affectation six months ago. Now, at the end of this conversation, it feels heavier than it should.

He is about to become, for the first time in thirty years of professional life, genuinely accountable for his decisions.

What Plato Knew, What Kahneman Proved, and What Jim Is Beginning to Understand

There is a through line from Plato's cave to Daniel Kahneman's System 1 and System 2, and from both to the conversation that just took place in a conference room on the fourteenth floor of a health system's administrative building. The through line is this: humans are not naturally good at knowing what they do not know.

Plato's prisoners did not know they were watching shadows. They had never seen the objects themselves. Their certainty about the shadows was absolute and entirely based on a limited experience that they had no framework for questioning. The tragedy of the cave is not that the prisoners were stupid. It is that they were epistemically closed, incapable of recognizing the limits of their own knowledge without an external intervention.

Kahneman's System 1 is the cognitive equivalent of the prisoners. It processes rapidly, confidently, and fluently, and it is precisely that fluency that makes it dangerous. System 1 produces what psychologists call cognitive ease, the comfortable feeling of understanding that comes when a problem is processed smoothly. Cognitive ease is pleasant, and it correlates poorly with accuracy.

Herbert Simon, one of the founding theorists of Decision Sciences and one of the twentieth century's most influential social scientists, described this phenomenon as bounded rationality, the observation that human decision-making is rational only within the bounds of the information available to the decision-maker and the cognitive capacity available for processing it. Simon won the Nobel Prize in Economics in 1978 for this insight. Its implications for organizational design are still being worked out.

The moleskin is a small tool for opening the cave. Writing down decisions, creating an external record that the decision-maker must later revisit and evaluate, is a cognitive intervention with a surprisingly long history. The Stoic philosophers of ancient Rome kept daily journals in which they recorded their decisions and examined their reasoning.

Marcus Aurelius' Meditations, which he wrote for his own use rather than for publication, is among the most detailed records of a leader's decision-making process that has survived from antiquity. Benjamin Franklin kept a systematic ledger of his decisions and their outcomes. Darwin kept field notebooks that functioned, among other things, as decision records, recording his observations, his interpretations, and his evolving conclusions.

The moleskin is not a new idea. It is a very old idea, rediscovered by every generation of thinkers serious enough about their own reasoning to subject it to external examination. What is new is the technology that allows us to scale that examination from individual reflection to organizational intelligence, to build, from thousands of individual decision records, a collective learning system that compounds in value over time.

That is the promise Susan has made to Jim. It is not a small promise.

"The beginning of wisdom," Aristotle wrote, *"is the definition of terms."* Jim has just learned to define decisions. This is not a trivial beginning. It is the precise beginning that every genuine inquiry requires, the moment when vague, comfortable, familiar words are replaced by sharp, demanding, accountable ones. Everything that follows depends on how well the definition holds.

Gerald's Story: The Decision That Made Jim an Executive

Remember Gerald, the patient who Jim couldn't diagnose properly and that one incident led to him becoming an C-level executive wanting to change how the system worked? Well, his story is a bit more nuanced than the version Jim tells everyone in the presentation and if you really want to understand decision sciences, this story is worth it...

Gerald was fifty-three when he first appeared in Jim's patient panel, referred by his previous physician who had retired. He was a high school history teacher, earnest, thoughtful, the kind of man who showed up to appointments prepared, with a list of questions written in a small spiral notebook. His hypertension had been diagnosed four years earlier. His type 2 diabetes was newer, discovered in an annual blood test two years prior.

Jim liked Gerald. It's common for physicians to develop a certain fondness towards their patients when both of them have active concerns and make active efforts towards health improvements. But, Jim's connection to Gerald was something more, something deeper.

Gerald reminded Jim of his father, someone who worked as an English teacher before retiring. With Gerald, there was something in his deliberateness, the way he responded to information, and his willingness to think that Jim naturally responded to.

While many might say this is a good thing, this fondness, as Jim would later reveal, was what was part of the problem.

You see, Jim was running late during the session where Gerald had mentioned fatigue and mild shortness of breath. The clinic was already having a tough morning because of an unexpected no-show that created a scheduling cascade that left Jim fifteen minutes behind by nine-thirty. So you see, he had been managing his way back toward the schedule all morning, shaving minutes from each appointment, and by the time Gerald arrived at eleven-forty-five, Jim had recovered eight of the fifteen minutes and was almost back on track.

Gerald's complaints, fatigue and mild shortness of breath, were, in Jim's initial pattern recognition and they were consistent with his existing notes:.

- Type 2 diabetes causes fatigue.

- His glucose control had been suboptimal for three months.

- The shortness of breath was mild enough to be attributable to the twenty pounds Gerald had gained over the past year, which was also contributing to his blood pressure elevation.

With this, Jim updated Gerald's medication, discussed diet and activity, and made a note to order a hemoglobin A1c test at the next visit.

What he did not do or what he would not have time to do in the time they spent together, was the calculation that would have changed everything. Gerald's age (sixty-three), his gender (male), his history of hypertension and diabetes, his weight, and the new-onset shortness of breath together produced a Framingham Heart Study risk score that, had Jim calculated it, would have been high enough to justify immediate further workup rather than routine follow-up.

But, Jim did not calculate it. He felt the visit was under control. He was running behind. Gerald's case pattern was familiar, diabetes and hypertension, metabolic management, nothing dramatic. The availability heuristic and the confirmation bias were working simultaneously, and Jim's System 2 was cognitively exhausted after three hours of continuous patient contact.

Gerald left the clinic at twelve minutes past noon. At eleven-fourteen the following morning, his wife called 911. The paramedics found him in the kitchen, pale and diaphoretic, with an anterior ST-elevation myocardial infarction.

Gerald survived. He recovered well enough to return to teaching after four months of cardiac rehabilitation. He sent Jim a note afterward, careful, thoughtful, written in the same deliberate hand as his appointment notebooks, that did not blame Jim but thanked him for his care over the years.

Jim read that note four times. Then he sat in his office for two hours and wrote, for the first time, what would become his personal philosophy of decision-making. He wrote that the:

- Quality of a decision cannot be evaluated by its outcome alone; that

- Cognitive conditions matter enormously and can be designed for or against; that

- Most dangerous moment in clinical decision-making is the moment that feels most familiar and routine; and that

- The difference between a good system and a good physician is that a good system catches what a good physician misses when the conditions are against them.

The note from Gerald still sits on Jim's desk and this personal philosophy? It will be in the first page of the moleskin Susan will hand him twenty years later. The two artifacts, the patient's gratitude, the physician's self-examination, are what will shape the intellectual journey that will go on to become MegaHealth's Decision Sciences program.

The Architecture of the Relationship: What Jim and Susan Are to Each Other

Okay so before we get into MegaHealth and what's going on in the entire healthcare organization when it comes to healthcare, it's very important to understand the bond Jim and Susan have.

Why?

Well, because that bond is exactly what this entire project is based on and it's essential to what keeps it running.

You see, they're not exactly friends, the power gap is too large, and their relationship remains quite professional, so that word doesn't quite fit. They aren't simply superior and subordinate, even though that formal relationship exists and is acknowledged by both.

While this connection between the two might be one that's hard to pinpoint, let's just say that they're two intellectual partners who have skills and blind spots that complement each other. With Jim you have:

- Domain expertise so deep that it constitutes genuine clinical wisdom.

- Organizational authority sufficient to make anything that is decided actually happen.

- A physician's diagnostic discipline when he is operating at his best.

- An honesty about his own limitations that is rarer in CEOs

But, he's not an expert at everything. He's not immune. Some of his key blind spots include:

- Specific overconfidence of being an experienced leader.

- A tendency to trust System 1 in domains where System 1 is not well-calibrated.

- A preference for decisive action that sometimes shortchanges the diagnostic pause.

- A natural authority that suppresses the dissent he needs.

Now, let's look at Susan. She has:

- A technical mastery of the behavioral science and decision theory literature that Jim engages with at a practitioner rather than expert level.

- A scientist's instinct for empirical rigor and discomfort with unexamined claims.

- The clarity and intellect to see patterns that others would miss.

However, some of her key blind spots include:

- A difficulty with the organizational politics

- An analytical perfectionism which delays good decisions in urgent scenarios.

- Occasional failure to translate insights into accessible, story-driven formats

So, you see, each of them compensates for the other's limitations in ways that neither acknowledges explicitly but that both have come to depend on. Jim tests Susan's analytical conclusions against organizational and political reality. Susan tests Jim's judgment calls against empirical evidence.

With this in mind, we can see that the coin flip and the Rock, Paper, Scissors game are not merely teaching tools. They are the opening moves in a dialogue that will continue for years, a dialogue in which both participants are learning, both are being changed, and both are building something that neither could build alone.

The Night Before: What Kept Jim Awake

The evening before the Tuesday morning meeting where Susan would hand him the coin, Jim drove home from the hospital through a light rain. He had been in a meeting from four to seven, a three-hour session with a consulting team from one of the major firms, presenting the findings of a six-month study on MegaHealth's strategic positioning.

The deck had been exhaustive. Ninety-two slides. Waterfall charts and bubble graphs and competitive landscape maps that stretched across three screens simultaneously. It had cost, by Jim's estimate, somewhere north of $800,000.

He had sat through the entire presentation with the attentive, engaged expression of a man who was listening carefully, because he was. And then, driving home through the rain, he tried to remember the three most important things he had learned.

He could not produce three. He could produce, with some effort, just one and that was MegaHealth's market share in cardiovascular services was declining modestly relative to two regional competitors. He had known this already. He had known about it for two years. The consulting team had quantified it more precisely and dressed the finding in more sophisticated analytical clothing, but the substance was not new.

Eight hundred thousand dollars for a fact he already knew.

This was not the consultants' fault. They had done the work honestly and competently. The problem was something upstream: the question they had been asked was not the right question, and no one, including Jim, had noticed until the answer was on the table and the invoice was in the mail.

The right question was not *"what is our competitive position?"* Jim knew his competitive position. He had been watching it for years. The right question should have been:

"Given our competitive position, what decision should we make, and how do we know if we're making it well?"

That question was never asked. It was assumed that the analysis would speak for itself, that the data would generate its own implications, that ninety-two slides of strategic intelligence would somehow convert themselves into organizational action.

They would not. They never did. That was what was keeping Jim awake.

The Physician's Diagnostic Mind: How Jim Actually Thinks

What Jim had brought from medicine to executive leadership, and what most physicians-turned-executives bring, whether they recognize it or not, is a diagnostic cognitive style.

Physicians are trained to move from symptoms to causes, to construct differential diagnoses, to weigh probability and update continuously as new evidence arrives. They are trained in a kind of Bayesian reasoning, starting with a prior probability (based on the base rate of the disease in the relevant population) and updating it as each piece of evidence arrives, that is, when it is functioning well, exactly the kind of reasoning that Decision Sciences prescribes.

The problem is that the diagnostic mind, transported from the exam room to the executive suite, encounters a domain where the feedback loops are radically different. In medicine, you make a diagnosis, prescribe a treatment, and learn the result within days or weeks. If you were wrong, you know quickly.

The feedback calibrates your diagnostic instincts over thousands of cases, building the reliable pattern recognition that makes experienced clinicians so effective.

In executive strategy, you pick a path, act on it, and wait years for the outcome. If you messed up, you might not realize it until the world has completely moved on. The gap is so wide that you don't actually learn from your mistakes. You could repeat the same blunder for ten years because the world never gives you a clear enough sign to stop.

Jim understood this intellectually but had not built a systematic response to it. He had been applying his diagnostic instincts, the rapid pattern recognition, the clinical intuition, the confidence in his own judgment, to a domain where those instincts were inadequately calibrated. He was using a well-sharpened tool in the wrong material.

The moleskin that Susan would hand him the next morning was, in one sense, the beginning of a new calibration exercise. He was going to take the diagnostic discipline he had built over thirty years of clinical training and apply it, deliberately and systematically, to the strategic decisions of organizational leadership. He did not know this yet. He would.

What Kahneman Learned from Meehl: The Actuarial Principle

One of the most unsettling findings in the behavioral decision literature, one that Daniel Kahneman describes with characteristic candor in Thinking, Fast and Slow, is the work of Paul Meehl, a Minnesota psychologist who published a book in 1954 with the innocuous title Clinical versus Statistical Prediction.

Meehl compared the predictive accuracy of trained clinical psychologists and psychiatrists making judgments about patient outcomes against simple statistical models that used the same information as the clinicians but processed it algorithmically.

The result was, and still is, really unsettling for anyone who trusts human expertise: in almost every study Meehl looked at, the math was as good as or better than the pro. Not just a few times, but basically every time. The expert's opinion, built on years of training, practice, and a *gut feeling* about the patient, was consistently worse than a simple formula that had never even met the person.

Meehl's finding has been replicated in dozens of domains since 1954. These include:

- Graduate school admissions.

- Parole decisions.

- Medical diagnosis.

- Financial forecasting.

- Wine quality prediction.

In each domain, simple actuarial models, which do nothing but apply consistent weights to a small number of relevant variables, outperform expert judgment. The reason, as Kahneman explains, is not that experts lack knowledge. It is that they apply their knowledge inconsistently, weighting variables differently in different cases, allowing irrelevant contextual factors to influence their judgments, and generally generating more noise than signal in the combination process.

The implication for organizational decision-making is stark and uncomfortable: for many categories of decisions, a well-designed decision rule or scoring model will outperform the expert judgment of the organization's most experienced and capable leaders. This is not a license for abandoning human judgment.

There are important decision categories, novel situations, decisions with complex ethical dimensions, decisions requiring genuine creativity, where algorithmic approaches are inadequate. But for the large category of structured, recurring decisions where the relevant variables can be identified and weighted, the evidence strongly favors the model over the expert.

Susan's decision scoring models, the weighted matrices she builds for capital allocation and vendor selection, are not bureaucratic formalities. They are actuarial tools designed to capture the signal in expert judgment while eliminating the noise. They are Meehl's insight, operationalized.

In Thinking, Fast and Slow, Kahneman describes what he considers the most personally important discovery of his career: the discovery, while serving as a young psychologist in the Israeli Army, that he and his colleagues were making predictions about the leadership potential of officer candidates with high confidence and virtually zero validity.

Despite receiving clear and consistent statistical feedback that their predictions were essentially uncorrelated with actual officer performance, they continued to make their assessments with undiminished confidence. Kahneman calls this the "illusion of validity", the subjective sense of certainty that accompanies an intuitive judgment, regardless of its actual accuracy.

The illusion is maintained by the coherence of the story the expert constructs: each piece of evidence fits neatly into a narrative that feels conclusive, even when it predicts nothing. This finding, which took Kahneman years to accept about himself, remains one of the most important in all of decision science: the feeling of confidence is not evidence of accuracy.

The decision process, not the feeling of the decision-maker, is the reliable indicator of decision quality.

Susan's Origin: What a Prediction Market Taught Her About Organizational Intelligence

Okay so now that we're back from that literature detour on behavioral decisions, let's look at Susan's background.

In 2007, she attended a conference on prediction markets at a business school in Cambridge. She was there because a friend from work told her it would be good for what she was doing with aggregating distributed information across large organizations, and it was. Susan didn't know this at the time, but this conference is what changed her career trajectory over time.

Prediction markets are information aggregation mechanisms that use financial market structure, buying and selling of contracts whose payoffs depend on the occurrence of specific events, to aggregate dispersed information into collective probability estimates. These markets were first used commercially by the Iowa Electronic Markets in the 1980s, and they gained widespread attention in 2004 when James Surowiecki published The Wisdom of Crowds, documenting the surprising accuracy of collective judgments relative to individual expert judgment.

The finding that electrified Susan was not the accuracy of prediction markets per se. It was what they revealed about the location of knowledge in organizations. In multiple corporate experiments with internal prediction markets, the employees who held the most accurate information about project outcomes, customer behavior, and market developments were not the senior executives making the official forecasts.

They were the frontline employees, the product managers, the sales representatives, the customer service agents., They were the ones who had direct contact with the reality the organization was trying to predict. The official forecasts were systematically wrong in predictable ways. They were:

- Optimistic.

- Anchored to strategic objectives.

- Insensitive to contrary field evidence.

The prediction market prices, which reflected the views of hundreds of employees rather than a small executive team, were consistently more accurate.

Susan's conclusion, which she spent the next decade developing into a practical framework, was that organizational intelligence is not concentrated in the leadership team. It is distributed across the organization, disproportionately located where the organization meets reality, in clinical practice, in customer interaction, in operational execution.

The job of an organizational Decision Sciences system is not to give the leadership team better tools for making decisions. It is to create mechanisms by which the distributed intelligence of the entire organization flows into the decision process. The leadership team's role is to set direction, establish governance, and ensure accountability, not to be the primary source of strategic intelligence.

This is why Susan's decision system at MegaHealth tracks decisions across all levels of the organization rather than just at the executive level. The 100,000 decisions logged in the first month are not an interesting statistic. They are the raw material of organizational intelligence that no leadership team, working from the top down, could ever generate on its own.

CHAPTER TWO: THE HIDDEN ARCHITECTURE OF CHOICE

Behavioral economics, cognitive bias, and the uncomfortable truth about how humans decide

Descartes' Error: The Myth of the Rational Mind

In the winter of 1619, René Descartes shut himself in a small, heated room in Germany and spent several days in intensive philosophical reflection. The three dreams he reported having during this retreat, which he interpreted as divine signs authorizing his philosophical project, launched one of the most consequential intellectual programs in Western history.

Descartes' ambition was nothing less than basing all human knowledge on a foundation of pure reason, purged of the errors of the senses and the confusions of received tradition.

The principle he arrived at, cogito ergo sum, *"I think therefore I am"*, placed rational thought at the very center of human existence and human identity. The mind and body, for Descartes, were separate substances. Reason was pure, reliable, and independent of the messiness of physical sensation and emotion. The rational mind was the locus of the human good.

For three hundred years, this idea from Descartes guided Western thought, science, and business. The notion of a logical person, someone who looks at data fairly, thinks through every choice, and picks the most useful option, became the starting point for all modern economics, management styles, and how we analyze decisions today. It is a beautiful vision. It is also, as the neuroscientist Antonio Damasio demonstrated in his landmark 1994 book *Descartes' Error*, fundamentally incorrect.

Damasio's central case involved a patient he called Elliot, a successful businessman who, following surgery to remove a brain tumor in his frontal lobe, emerged with his IQ, memory, and analytical capabilities entirely intact but his decision-making catastrophically impaired.

Elliot could not make simple choices. He would spend forty-five minutes deliberating over where to have lunch. He could analyze options indefinitely but could not commit. He had lost, through the damage to his ventromedial prefrontal cortex, the emotional machinery that normally provides the *"somatic marker"*, the embodied feeling of rightness or wrongness, that guides decision-making.

Damasio's conclusion was radical and uncomfortable: emotion is not the enemy of good decision-making. Emotion is an essential component of it. A brain without emotional input cannot decide effectively, because it cannot determine which considerations are worth caring about. The rational actor model, the Cartesian executive who weighs evidence and chooses optimally, is not just empirically wrong. It is neurologically impossible.

This is huge for Decision Science. It means the goal isn't to kick emotion out of the office, but to see how feelings and facts mix. We need to find where our emotions trip us up and build a better way to make decisions, one that fixes those mistakes without pretending we can just turn off the human side of making a choice.

The Somatic Marker Hypothesis

Antonio Damasio's Somatic Marker Hypothesis proposes that emotional signals, experienced as physical sensations, or "somatic markers", bias cognitive processing toward or away from certain decision options, effectively pre-screening the option space before conscious deliberation begins.

This process is normally adaptive: extensive experience produces reliable emotional signals that guide decision-making efficiently in familiar domains. The dysfunction occurs when the emotional signal is calibrated to the wrong reference experience, when, for example, a leader's past success with a particular strategy generates an emotional signal of confidence that persists even when current conditions have made the strategy inappropriate.

Training Decision Scientists to recognize the difference between reliable somatic markers and mis-calibrated ones is one of the central practical challenges of the field.

Two Psychologists Who Changed Economics Forever

In 1969, a young Israeli psychologist named Amos Tversky walked into the office of his colleague Daniel Kahneman at the Hebrew University of Jerusalem and proposed a collaboration. Tversky was dazzling and supremely confident; Kahneman was thoughtful and self-questioning.

Together, they would produce a body of work that fundamentally altered the scientific understanding of human judgment and decision-making and eventually earned Kahneman the Nobel Prize in Economics in 2002 (Tversky had died in 1996, and the Nobel is not awarded posthumously).

They started working together after noticing something simple: even experts who know math inside and out get probability wrong in the exact same ways. Even professors, when faced with certain problems, made mistakes that broke the very rules they taught. It wasn't because they were uneducated. It was something more fascinating and a bit scary: our brains use mental shortcuts that, under the right conditions, lead us into the same traps every single time.

Kahneman and Tversky's program of research over the next two decades identified and documented dozens of such heuristics and the biases they produce. Their framework, eventually formalized in Kahneman's 2011 book Thinking, Fast and Slow, distinguishes two cognitive systems.

System 1 and System 2: The Architecture of Cognition

To understand how we make choices, we must look at the two engines that drive our thinking. Most of the time, we imagine ourselves as logical people making careful plans, but the reality is much more complicated.

Our brains use two completely different systems to process information, and they are constantly in a tug-of-war. Understanding the difference between them is the only way to figure out why smart people make big mistakes, especially when the pressure is on.

System 1 is the brain's fast, automatic, intuitive mode of processing. It operates continuously and largely below conscious awareness, drawing on pattern recognition, emotional associations, and learned heuristics. This is what generates rapid judgments and responses.

It allows a chess grandmaster to recognize a favorable board position at a glance, a cardiologist to sense that a patient *"doesn't look right"* before any diagnostic data is reviewed, and an experienced executive to read a room in the first thirty seconds of a meeting.

You can think of this as a repository of genuine expertise, and its outputs, when the domain is familiar and the feedback has been reliable, are often superior to deliberate analysis.

System 2, on the other hand, is the brain's slow, deliberate, analytical mode. It handles logical reasoning, arithmetic, hypothesis testing, and the construction of novel arguments.

This system can override System 1's outputs, but only when it is engaged, which requires motivation, cognitive resources, and time. System 2 is expensive: it depletes glucose, creates what Baumeister calls *"ego depletion,"* and degrades under conditions of time pressure, cognitive load, or emotional stress.

These conditions, time pressure, cognitive load, emotional stress, are the defining features of most high-stakes organizational decisions.

The takeaway for any business here is quite useful: the exact moments when we need to be most careful, when things are urgent, complex, and stressful, are the same moments our logical brain shuts down and our gut takes over. If we don't intentionally design a better way to make decisions, our biggest moves will be handled by our least reliable way of thinking.

> *"It is an important fact about our minds that cognitive ease makes us feel good. A sentence that is printed in clear type, or has been repeated, or has been made rhyming, will be judged as more true than one that is difficult to read, novel, or cacophonous. And this holds whether or not the statement is actually true."*
>
> **Daniel Kahneman, Thinking, Fast and Slow**

The Bias Catalog: What We Know About How We Go Wrong

Researchers have uncovered more than 180 cognitive biases, the mental *"glitches"* that skew our perception of reality. For healthcare leaders, a handful of these are particularly risky. Rather than drowning you in theory, this section will focus on the specific biases most likely to cloud your judgment during high-stakes clinical and business decisions.

1. Anchoring and Adjustment

In 1974, Kahneman and Tversky published a landmark paper titled *"Judgment Under Uncertainty: Heuristics and Biases"* in Science. One of the most powerful traps they discovered is *"anchoring."* This is our tendency to latch onto the first piece of information we hear and let it weigh down every decision we make after that. The scary part? This happens even when that first bit of info is random or has nothing to do with the problem at hand.

In their original experiments, participants spun a wheel of fortune that landed on a random number, then were asked to estimate the percentage of African countries in the United Nations. Participants who spun high numbers (65) gave median estimates of 45 percent. Participants who spun low numbers (10) gave median estimates of 25 percent. The wheel, which was known by all participants to be random, had anchored their estimates.

In healthcare strategy, anchoring manifests in ways that are both pervasive and consequential. When a vendor presents a comprehensive $50 million solution before revealing a $12 million phased alternative, the $50 million anchors all subsequent financial evaluation.

When a board member mentions that a competitor is expanding into three new markets, the number three anchors subsequent discussion of MegaHealth's expansion plans. When a consultant presents a range of outcomes ($3M to $20M), the midpoint ($11.5M) anchors the committee's base case estimate.

The *"de-anchoring"* technique is one of the most powerful tools in a leader's kit. It involves a few simple steps: looking at the data before hearing any initial numbers, having team members come up with their own estimates privately before talking as a group, and calling out the risk of anchoring as soon as the meeting starts. These strategies help ensure the first number mentioned doesn't end up controlling the entire conversation.

When I run Community Health Needs Assessments, I always give a data presentation to begin the day, before the group starts the strategy work. It is important to be grounded in reality. This also helps to minimize the baggage that many Directors of a service line bring to the table.

We can see exactly how this works in practice by looking at what Susan had to say to Jim during their Linac preparation. *"Before we meet with the vendors, I want you to write down your internal estimate of what a single linear accelerator should cost, based on your general knowledge of medical equipment. Write it down and don't share it yet. When the vendors present their numbers, you'll have a pre-anchor reference point that's genuinely yours, not theirs."*

2. Prospect Theory and Loss Aversion

Kahneman and Tversky's Prospect Theory, published in 1979 in Econometrica, is among the most cited papers in the history of economics. It challenged the foundational assumption of expected utility theory, that people evaluate outcomes in terms of their absolute value, by demonstrating that people evaluate outcomes in terms of their deviation from a reference point, and that losses from that reference point loom approximately twice as large psychologically as equivalent gains.

The theory also identified two other consistent features of human valuation: diminishing sensitivity (the difference between losing $100 and $200 feels smaller than the difference between losing $0 and $100, even though both are $100 differences), and probability weighting (people overweight small probabilities and underweight large ones, explaining both the appeal of lotteries and the purchase of insurance for low-probability catastrophes).

For healthcare executives, loss aversion has three specific and costly manifestations. The first is status quo bias, the tendency to prefer doing nothing over taking action, because action creates the risk of loss while inaction preserves the current reference point. The second is escalation of commitment, the tendency to persist in losing strategies because abandoning them requires consciously realizing a loss.

The third is reference-point dependency in financial framing, the tendency to evaluate the same financial outcome differently depending on whether it is presented as a gain relative to a lower baseline or as a loss relative to a higher baseline.

"Losses loom larger than gains", Kahneman and Tversky, Prospect Theory (1979)

This six-word summary has generated more economic research than perhaps any other statement in the discipline's history. Its practical implications for healthcare capital allocation are still not fully appreciated.

Consider the standard committee discussion of a capital investment. The financial model shows a base-case NPV of $5.7 million over five years. Loss aversion predicts that the committee will psychologically weigh the downside scenario (NPV of $1.2 million, or a complete write-off of the investment) approximately twice as heavily as the upside scenario (NPV of $9.3 million).

The rational expected-value calculation favors investment; the loss-averse psychological evaluation may not. And the committee members, absent explicit Design Science intervention, will not be aware of this asymmetry in their own processing.

The Framing Effect: Same Math, Different Decisions

In a classic experiment by Kahneman and Tversky, participants were asked to choose between two public health programs responding to a disease expected to kill 600 people.

In the "gain frame," Program A would "save 200 people" with certainty; Program B offered a 1/3 chance of saving all 600 and a 2/3 chance of saving no one. In the "loss frame," Program A would "result in 400 deaths" with certainty; Program B offered a 1/3 chance that nobody dies and a 2/3 chance that all 600 die. The math is identical in both framings.

But when framed as gains, 72 percent of participants chose the certain option (Program A). When framed as losses, 78 percent chose the gamble (Program B).

The framing of the same decision, not its content, changed the majority preference. Healthcare policy, clinical guideline communication, and strategic plan presentations are all subject to this effect, continuously and largely without awareness.

3. Confirmation Bias and the Sacred Assumption

Confirmation bias, the tendency to seek, interpret, and remember information in ways that confirm pre-existing beliefs, may be the single most consequential cognitive error in organizational life. It is pervasive, subtle, and self-reinforcing: the more confidently an organization believes something, the more effectively it filters out evidence to the contrary.

The philosopher Francis Bacon described confirmation bias in 1620, in the Novum Organum, as one of the *"Idols of the Tribe"*, systematic errors inherent in human nature rather than individual character. *"The human understanding,"* Bacon wrote, *"when it has once adopted an opinion, draws all things else to support and agree with it."* Bacon was writing about scientific inquiry, but his diagnosis applies with equal force to strategic planning, competitive analysis, and clinical decision-making.

In healthcare organizations, confirmation bias manifests at multiple levels. At the individual level, physicians who form an early diagnostic hypothesis attend more closely to confirming evidence and rationalize disconfirming evidence, a phenomenon called premature closure that contributes to a significant proportion of diagnostic errors. At the organizational level, leadership teams that have committed to a strategic direction seek data that supports it, dismiss data that challenges it, and attribute failures to implementation rather than strategy.

The most dangerous form of confirmation bias in organizational settings is what psychologist Irving Janis called groupthink, the convergence of a cohesive group around shared beliefs that are never seriously challenged because the social cost of dissent exceeds its perceived value. Janis developed his groupthink concept by analyzing several catastrophic decision failures, including the Bay of Pigs invasion, the escalation of the Vietnam War, and the failure to anticipate the attack on Pearl Harbor.

In each case, highly capable decision-makers, operating with substantial information, reached conclusions that most dispassionate observers would have recognized as badly wrong. The failure was not analytical capacity. It was the social and psychological architecture of the decision environment.

This structural failure isn't just a theoretical problem; it shows up in the way teams talk to each other every day. Often, it takes an outside perspective to see how these patterns of thought are actually playing out in real time. Susan saw this happening during Jim's meetings and decided to point out the specific habit that was sabotaging their logic.

She said, *"Jim, I have observed something in your last four strategy sessions that I want to name directly. When the data supports the direction the room is already leaning, it is accepted without question. When it challenges the prevailing view, someone, always someone different, but always someone, finds a reason to question the methodology, the sample, or the source. This happens so consistently that it looks less like genuine analytical skepticism and more like a social mechanism for protecting sacred assumptions. That is confirmation bias operating at the organizational level. And it means that your best data is systematically less influential than your worst instincts."*

To this, Jim replied, *"That's... a very uncomfortable observation."*

"All the useful ones are," said Susan.

4. The Availability Heuristic and the Distortion of Probability

The availability heuristic, identified by Kahneman and Tversky in 1973, describes the human tendency to judge the probability of an event by how easily examples come to mind. Events that are vivid, recent, or emotionally resonant seem more probable than events that are abstract, distant, or emotionally neutral, regardless of their actual base rates.

After the Ebola outbreak of 2014, hospital systems across the United States invested heavily in Ebola preparedness protocols that, in retrospect, represented a significant misallocation of infection control resources. The vividness of the Ebola imagery in the media, the hazmat suits, the isolation chambers, the dramatic reporting, made Ebola seem far more likely to affect the average American hospital than any actuarial analysis supported.

Meanwhile, far more probable and immediately harmful threats, influenza, MRSA, Clostridium difficile, received relatively less attention because they were familiar, undramatic, and cognitively *"available"* primarily to specialists rather than to the general public and administrative leadership.

For healthcare executives, the availability heuristic distorts risk assessment in both directions. Recent crises are overweighted; chronic, undramatic risks are underweighted. The experience of a specific failure, a regulatory penalty, a high-profile readmission, a nurse retention crisis, calibrates risk perception to that failure for months or years afterward, distorting resource allocation toward the prevention of the specific event that already occurred.

The corrective is base-rate thinking, the deliberate practice of comparing any proposed risk or opportunity against the historical frequency of similar events, rather than allowing recent or vivid instances to dominate the probability estimate. This is the core idea behind "reference class forecasting," which we'll cover in the next chapter.

5. The Planning Fallacy and Optimism Bias

Daniel Kahneman and Amos Tversky described the planning fallacy in 1979: the systematic tendency to underestimate the time, costs, and risks of planned projects while simultaneously overestimating the benefits. Everyone falls into the "planning fallacy." It happens to individuals, big companies, and even governments. Interestingly, experience doesn't seem to help; even people who have managed dozens of projects still tend to guess that future ones will be cheaper and faster than they actually turn out to be.

The planning fallacy is driven by what Kahneman calls the inside view: the tendency to focus on the specific details of a project when estimating its timeline and cost, rather than on the base rate of similar projects. When you think about building a new cardiology clinic, you think about this specific clinic, its specific location, its specific design, its specific physician team.

You do not naturally think about the historical average time and cost of building cardiology clinics in your market, adjusted for the base rate of permitting delays, construction cost overruns, equipment procurement delays, and physician recruitment misses. But it is the historical base rate, not the optimistic specifics of your inside view, that most accurately predicts your future experience.

The antidote to the planning fallacy is what Kahneman calls the outside view, the deliberate practice of identifying a reference class of past projects similar to the one being planned, examining the distribution of actual outcomes in that reference class, and using those base rates to calibrate the current estimate.

This practice, called reference class forecasting, was first formally developed by the economist Bent Flyvbjerg at Oxford's Saïd Business School and has been shown to dramatically improve forecast accuracy for large infrastructure projects. It is directly applicable to healthcare capital planning.

6. Overconfidence: The Most Dangerous Bias in Healthcare Leadership

Of all the documented cognitive biases, overconfidence may be the single most consequential for organizational decision-making. Overconfidence is the systematic tendency for stated confidence to exceed actual accuracy, for people to believe they know more, predict more accurately, and control outcomes more than the evidence supports.

The research on overconfidence is both extensive and humbling. In one classic study, participants asked factual questions were asked to provide a confidence interval, a range they were 90 percent certain contained the true answer. Across dozens of studies, participants' 90 percent confidence intervals captured the true answer only 50 to 70 percent of the time.

The excess confidence is not the result of ignorance; experts in specific domains show the same pattern, and in some studies show greater overconfidence than generalists, because their domain knowledge generates stronger feelings of certainty.

For healthcare executives, overconfidence manifests in three particularly costly ways that include:

- Strategy - leaders consistently overestimate their organization's competitive position, their implementation capacity, and the accuracy of their market projections.

- Recruitment - hiring managers overestimate their ability to predict job performance from interviews, a finding so robust that structured interview protocols, designed precisely to reduce the influence of interviewer confidence, consistently outperform unstructured interviews in predicting actual performance.

- Crisis Management - executives overestimate their ability to manage crises as they unfold, leading to under-preparation and over-confidence in improvised responses.

The coin flip exercise Susan runs with Jim is a direct intervention against overconfidence. Its power is not in the information it provides, Jim already knows that coin flips are random. Its power is in the direct, experiential confrontation with the gap between felt confidence and actual accuracy. No amount of explaining overconfidence produces the same effect as experiencing it in real time.

7. The Sunk Cost Fallacy and Escalation of Commitment

The sunk cost fallacy, the tendency to continue investing in a course of action because of past investments, rather than based on the prospective value of continued investment, is among the most economically costly cognitive errors in organizational life. The rational principle is simple: past costs are irretrievable. They should have no bearing on future decisions, which should be based solely on prospective costs and benefits.

But humans are not rational about sunk costs. We feel them. We are loss-averse, prospect theory tells us that, and abandoning a project in which we have invested heavily feels like consciously accepting a loss. The psychological pain of that loss motivates us to continue investing, even when the prospective returns no longer justify it, because continuing allows us to maintain the hope, however implausible, that the investment will eventually pay off.

In healthcare, sunk cost effects appear at every level: the health system that continues to operate a money-losing service line because it *"has always been part of our mission"*; the department that continues using an outdated software system because the original implementation cost $4 million; the clinical program that persists in an ineffective protocol because changing it would require acknowledging that the previous protocol was wrong. In each case, the relevant question, *"Does continuing this course of action create more value than the next best alternative use of these resources?"*, is displaced by the emotionally driven question: *"Can we afford to admit we were wrong?"*

The organizational corrective for sunk cost bias is what decision analysts call a *"sunset provision"*, a pre-committed decision rule specifying the conditions under which a project will be terminated, agreed upon before the project begins, when the sunk cost pressure is zero. Susan will build sunset provisions into MegaHealth's decision records for exactly this reason.

8. Authority Bias and the Stifling of Dissent

Authority bias, the tendency to attribute greater accuracy and wisdom to the views of authority figures, independent of the actual quality of their reasoning, is among the most pervasive and least discussed sources of organizational decision failure. It operates through several mechanisms that include:

- **Direct Deference** - Junior team members change their stated positions when they learn the senior person's view)

- **Selective Amplification** - Arguments aligned with the leader's position receive more attention and credence than those that contradict it

- **Self-censorship** - team members with contradicting information or views decide not to raise them, anticipating negative social consequences.

The management scholar Amy Edmondson at Harvard Business School has spent three decades studying what she calls psychological safety, the belief that one will not be punished or humiliated for speaking up with ideas, questions, concerns, or mistakes. Her research consistently shows that teams with higher psychological safety make better decisions, learn more effectively from errors, and produce better outcomes across a wide range of domains, including healthcare.

The hospitals with the highest error-reporting rates, which seem, counterintuitively, to be performing worse, actually have better patient safety records, because psychological safety creates an environment in which errors are surfaced and corrected rather than concealed and compounded.

For Jim, authority bias is a structural problem. His position as CEO with thirty years of professional experience and a reputation for sound judgment, creates exactly the authority gradient that stifles the dissent he most needs.

The techniques that Susan will introduce, structured devil's advocacy, pre-mortem analysis, anonymous input collection, are all designed to engineer psychological safety into the decision process, creating structural channels for dissent that bypass the authority gradient.

Richard Thaler and the Architecture of Better Choices

Many may not know this but the response to the behavioral economics literature, the question of knowing what to do with cognitive bias rather than just document it is something that Richard Thaler and Cass Sunstein do share in the book Nudge: Improving Decisions About Health, Wealth, and Happiness. Thaler, a Nobel Prize winner in Economics in 2017, developed the concept of the nudge.

According to him, it's a change in the choice architecture that happens when the way options are presented and the context in which decisions are made is altered. He believed that this change can influence one's behavior without forbidding any option or changing the economic incentives and that such an influence can be predictable as well.

The most famous example is default options. When organ donation rates across countries were first systematically compared, researchers discovered a startling pattern: countries that required active opt-out from organ donation (Germany, UK, US) had rates between 10 and 30 percent. Countries that required active opt-in to non-donation (France, Austria, Spain) had rates above 90 percent. The difference was not cultural, economic, or religious. It was purely structural: the same decision, framed as an active choice in two different directions, produced profoundly different outcomes because of the power of defaults.

Thaler called the designers of choice architecture "choice architects", a role that every organizational leader, every process designer, and every communication professional already plays, whether or not they recognize it. The question is not whether choice architecture exists; it is whether it is being designed intentionally or operating by default.

For Decision Science practitioners in healthcare, choice architecture operates at multiple levels that include:

- The order in which options are presented in strategy meetings

- The defaults built into clinical order sets.

- The framing of financial options in capital requests

- The design of decision support tools in electronic health records

- The structure of escalation processes in risk management.

Each of these is a choice architecture and each can be designed to work with, or against, the predictable tendencies of human cognition. Keep this in mind as Susan brings this week's framework to Jim and says:

> *"Last week you tracked your decisions in the moleskin. Let's review. How many significant decisions did you make in five days of work?"*

Jim, Counting. *"I wrote down twenty-three. But I think I missed some."*

Susan, *"How many were made in the first two hours of the workday versus the last two?"*

Jim pages through the notebook. He counts.

Jim, *"Fourteen in the morning. Nine in the afternoon."*

Susan, *"Look at the afternoon decisions. Read me a few."*

Jim reads. The afternoon entries are shorter, less specific, more conclusory. Three entries simply say *"agreed Mathew's recommendation."* Two say *"went with my gut.*" The morning entries are more deliberate, they note alternatives considered, they articulate reasons.

Jim, *"The afternoon entries are less thoughtful."*

Susan, *"That is decision fatigue. Your System 2 was depleted by the time you reached the afternoon. You were running on System 1, pattern matching, intuition, social signals, and whatever the room seemed to want. This is not a character flaw. It is a neurological constraint. But it is a constraint you can design around. High-stakes decisions should happen in the morning. Complex analysis reviews should be scheduled before lunch, not after. The most consequential choices in your organization's week should not be occurring at three in the afternoon after six hours of continuous cognitive load."*

Jim, *"I've scheduled a capital committee meeting for four-thirty on Fridays for the past three years."*

Susan, *"Then we are moving it to nine-thirty on Tuesdays."*

Dan Ariely and the Predictably Irrational Organization

While Kahneman and Tversky focused on the mechanics of individual thought, Dan Ariely, Duke University economist and author of Predictably Irrational, broadens the scope to include the social and organizational forces behind our mistakes.

Ariely argues that human choice is rarely a solitary act. He examines how social context, ranging from peer pressure and fairness norms to the symbolic value we place on our actions, steers behavior in ways that purely cognitive models miss. By highlighting these external architectures, Ariely proves that our irrationality is not just a mental glitch, but a product of the world around us.

One of Ariely's most provocative findings concerns what he calls the *"price zero"* effect: the disproportionate appeal of free options, which are preferred over merely cheap options even when the cheap option represents significantly greater value. Healthcare organizations are not immune to this effect. The *"free"* EHR module bundled with an existing contract, the *"free"* consulting hours offered by a vendor as part of a larger deal, the *"free"* continuing education provided by a pharmaceutical company, each carries hidden costs in the form of influenced decisions that the recipient does not consciously register.

Ariely has also documented what he calls the *"relativity trap"*, the tendency to evaluate options not on their absolute merits but relative to the other options present. The design of the choice set, which options are included, which are excluded, and how they are ordered and presented, systematically influences which option is selected, independent of its objective value. This is the principle behind the restaurant menu trick of including a very expensive option (the $120 prix fixe) that nobody orders but that makes the $65 second-tier option seem reasonable by comparison.

For healthcare strategy, the relativity trap appears in every request for proposal, every vendor presentation, and every capital allocation committee meeting. The options presented shape the decision more powerfully than the evaluation criteria applied to them.

When the Stakes Are Human Lives: Cognitive Bias in Clinical Settings

Everything discussed in this chapter applies with doubled force to clinical decision-making, where the consequences of cognitive error are measured not in dollars but in health outcomes and, sometimes, lives.

The most extensively documented form of clinical cognitive error is diagnostic error, the failure to reach the correct diagnosis in an appropriate time frame, resulting in harm to the patient.

A landmark systematic review by Mark Graber and colleagues identifies a critical imbalance in the mechanics of clinical failure. Analyzing 100 instances of diagnostic error, the study found that cognitive factors were implicated in 74% of cases, outperforming system-based failures, such as communication gaps or technical hurdles, by a factor of three.

The research highlights a specific "trinity" of mental traps that skew clinical reasoning:

- **Premature Closure:** The tendency to finalize a diagnosis before considering all viable alternatives.

- **Anchoring:** An over-reliance on the initial piece of information, which exerts a disproportionate influence on all subsequent reasoning.

- **Confirmation Bias:** A selective search for data that reinforces a starting hypothesis rather than objectively challenging it.

The neurologist Groopman, in his widely read 2007 book How Doctors Think, presents dozens of case studies illustrating these mechanisms in action. The common thread is not physician incompetence but cognitive overconfidence, the belief, usually implicit, that the physician's initial pattern-recognition response is accurate, and that further deliberate analysis is unnecessary.

As Groopman argues, the corrective is not more medical knowledge but better metacognitive habits: the disciplined practice of asking, for every diagnosis, *"What else could this be? What evidence would change my mind? When did I last update my hypothesis?"*

These metacognitive habits are precisely what Decision Sciences systematizes at the organizational level. The decision process architectures that Susan builds for MegaHealth are, in a deep sense, organizational metacognition, the institutionalization of the self-questioning habits that the best individual decision-makers practice informally but that most organizations have never explicitly designed.

Nudge Architecture in Clinical Settings

A series of landmark studies by Wachter, Bates, and colleagues at UCSF and Harvard demonstrated that clinical order set design, the architecture of options presented to physicians in electronic health records, dramatically influences prescribing behavior, test ordering, and protocol adherence.

In one study, changing the default antibiotic prescription from broad-spectrum to narrow-spectrum for common infections reduced inappropriate broad-spectrum prescribing by 47 percent, with no mandates, penalties, or direct clinician communication. The entire effect was achieved through architecture, changing the default, not the law.

This is the Thaler-Sunstein nudge applied to clinical Decision Sciences. The implications for hospital protocol design, medication administration systems, and surgical checklist architecture are profound and largely unexplored.

From Diagnosis to Architecture: The Decision Sciences Response

The catalog of cognitive biases could be demoralizing. If our cognition is this systematically flawed, if we anchor inappropriately, lose track of base rates, underestimate costs, overestimate our own accuracy, and systematically attend to confirming over disconfirming evidence, then how can we make good decisions at all?

The answer, and this is the central practical proposition of behavioral economics as a discipline, is that awareness of bias does not cure it, but architecture can compensate for it.

Despite a career spent cataloging the pitfalls of "System 1," Daniel Kahneman argues against its suppression. Attempting to operate exclusively through the analytical "System 2" is not only neurologically impossible but often counterproductive; in high-stakes domains of genuine expertise, System 1's pattern recognition frequently outperforms deliberate analysis.

The objective, therefore, is not the elimination of intuition, but the optimization of the decision-making process. This involves:

- **Selective Engagement:** Activating System 2 precisely when the stakes or complexity demand it.

- **Structural Interventions:** Designing "choice architectures" that safeguard against the most damaging biases.

- **Feedback Loops:** Implementing rigorous evaluation systems to improve mental calibration over time.

This is exactly what Susan is building at MegaHealth. Not a system that removes human judgment from decisions. A system that makes human judgment more accurate, more accountable, and more improvable. A system in which the cognitive errors documented in this chapter become organizational learning opportunities rather than organizational secrets.

Plato wanted to lead the prisoners out of the cave. Kahneman showed us why the cave is impossible to leave entirely. Thaler and Sunstein showed us how to redesign the cave so that the shadows on the wall are less systematically misleading. And Susan, working with

Jim at MegaHealth, is about to do exactly that.

"The best argument against democracy," Winston Churchill once observed, *"is a five-minute conversation with the average voter."* The best argument against human decision-making is five minutes with a behavioral economics literature review. But Churchill also said democracy is the worst form of government, except for all the others. Human decision-making is the worst cognitive process for managing complex organizations under uncertainty, except for all the alternatives.

The goal is not to find a substitute. It is to make it as good as it can be.

Philip Tetlock and the Superforecasters

In 2005, Philip Tetlock, a psychologist at Berkeley, published a twenty-year study of political forecasting that produced one of the most thorough and sobering assessments of expert judgment ever conducted. Tetlock had recruited 284 experts, political scientists, economists, policy analysts, intelligence officers, and journalists, and asked them to make nearly 28,000 predictions about political and economic events over the following two decades.

His finding, summarized in the deliberately provocative subtitle of his book Expert Political Judgment: How Good Is It? How Can We Know?, was that the average expert was slightly less accurate than a simple algorithm that predicted *"more of the same."*

The most interesting finding, however, was not the average. It was the variance. Some experts were reliably better than others. Not by much, but consistently. And the characteristic that distinguished the better forecasters from the worse was not domain expertise, political ideology, or academic credentials. It was what Tetlock called the *"fox versus hedgehog"* distinction, borrowed from the philosopher Isaiah Berlin's reading of the Greek poet Archilochus.

The hedgehog knows one big thing: a single organizing principle, an ideology, a theory, a framework, that it applies to every question. The fox knows many small things: it draws on multiple theories and frameworks, updates readily in response to new evidence, and is comfortable with uncertainty and contradiction. Hedgehogs make confident, memorable predictions, but foxes make ones that are more accurate..

In a follow-on project called the Good Judgment Project, funded partly by the Defense Advanced Research Projects Agency, Tetlock identified a small subset of participants who were dramatically more accurate than the average, people who, over thousands of forecasts on subjects ranging from geopolitics to economics to public health, consistently beat not just other forecasters but classified intelligence assessments from teams with access to far more information. These people he called superforecasters.

The characteristics of superforecasters are remarkably consistent with the prescriptions of Decision Sciences. They make probabilistic rather than binary predictions, not *"X will happen"* but *"X has a 67 percent probability of happening."* They update their probability estimates frequently as new information arrives. They are voracious consumers of base rate data. They are actively open to disconfirming evidence. They decompose complex questions into smaller, tractable sub-questions. And they are intensely self-critical about their track records, spending more time analyzing their errors than celebrating their successes.

The organizational implication is direct: the qualities that define a "superforecaster" are not innate gifts, but skills that can be trained and systematized. Practices such as probabilistic thinking, reference class reasoning, active updating, and honest error analysis are exactly that, practices.

When these habits are modeled by leadership and reinforced by corporate culture, they transition from individual traits to organizational assets. Given this, they provide a durable analytical advantage over less disciplined competitors. This shift from talent to skill is why Susan sees Tetlock's findings as the most actionable when it comes to modern decision making. Explaining what she thinks, Susan says to Jim:

"Tetlock's finding about superforecasters is, for me, the most practically important result in all of decision research. It says that the gap between a good forecaster and a great one is not raw intelligence. It is a specific set of cognitive habits, habits that, with the right feedback and the right culture, any intelligent person can develop. That means we can build a culture of super forecasting at MegaHealth. Not overnight. But over the years we have been working together, there is no reason why our senior leadership team cannot be making predictions, about market developments, about clinical outcomes, about competitive responses, at the level of Tetlock's best forecasters."

To this Jim replies, *"What would that look like in practice?"*

"It looks like what we are building. Decision Records that include explicit predictions. Outcome reviews that compare those predictions to what actually happened. A culture where being wrong is information rather than embarrassment. Probabilistic language in presentations, not 'this will work' but 'we estimate a 70 percent probability that this achieves its primary objective.' These are changes in language and habit that, compounded over years, produce the mental discipline of the superforecaster," explains Susan.

The Narrative Fallacy: Why We Are Addicted to Stories That Explain Too Much

The organizational implication is direct: the attributes of a *"superforecaster"* are trained practices, such as probabilistic thinking and active updating, rather than innate gifts. For Susan, this makes Tetlock's research the most actionable insight for MegaHealth, suggesting that leadership can close the performance gap by cultivating specific cognitive habits rather than relying on raw intelligence.

However, achieving this requires confronting Kahneman's narrative fallacy, the seductive human tendency to mistake a compelling causal story for factual evidence. By institutionalizing feedback and rewarding these habits, an organization can dismantle these neat, misleading narratives and replace them with a durable, data-driven analytical advantage.

The narrative fallacy is not a defect of *"bad thinkers,"* but a fundamental feature of the human cognitive apparatus, deeply embedded in the architecture of memory and communication. As story-processing creatures, we instinctively organize experience into a protagonist-led arc of goals, obstacles, and consequences to make information memorable and transmissible.

While this structure makes stories the most powerful communication format in human culture, it also creates a strategic vulnerability: the more *"readable"* and coherent a story is, the more likely we are to accept it as an accurate causal account, often ignoring the role of randomness or conflicting data in the process.

The problem arises when the narrative's compellingness is interpreted as evidence of its accuracy. A strategic retrospective that attributes an organization's success to the wisdom and vision of its leadership is almost always more accurate than an alternative account that attributes the success primarily to favorable market conditions and fortunate timing, not because the leadership account is truer but because it is a better story.

It has a clear protagonist, a coherent arc, and a satisfying causal connection between decision and outcome. The luck account is diffuse, attributional unsatisfying, and feels like it denies the importance of human agency.

Nassim Nicholas Taleb, in his 2007 book The Black Swan, named this phenomenon the *"hindsight bias amplifier"*, the way that causal narratives constructed after the fact systematically overstate the predictability of outcomes that were, before the fact, genuinely uncertain.

Jim Collins' Good to Great, Michael Lewis's Moneyball, and dozens of other celebrated business narratives are vulnerable to this critique: they construct compelling causal stories from data selected to support a predetermined conclusion, and the stories feel true because they are well-told, not because the causal identification is rigorous.

Susan has a specific protocol for addressing the narrative fallacy in MegaHealth's strategic retrospectives. Before any post-hoc analysis of a major decision's outcome, she requires the team to construct two competing narratives: one in which the decision caused the outcome, and one in which the outcome would have occurred regardless of the decision.

Both narratives are presented to the group before the retrospective analysis begins. This forces explicit consideration of the counterfactual, the *"what would have happened anyway"* question that the narrative fallacy systematically suppresses.

Gary Klein, Naturalistic Decision Making, and When Experts Are Right

The literature of decision science can sometimes feel like an extended argument against human judgment, a systematic indictment of the cognitive processes that experienced professionals have spent careers developing. This is a misreading that Gary Klein, a cognitive psychologist who founded the field of Naturalistic Decision Making, has spent three decades correcting.

He began his research program in the 1980s with a study of firefighting, specifically, how experienced fire commanders made decisions in the field under conditions of extreme time pressure, high stakes, and limited information. The received wisdom, based on laboratory decision research, predicted that commanders would engage in comparative analysis: generating multiple options and selecting the best one. Klein found something completely different. Experienced commanders almost never generated multiple options. They recognized the situation as belonging to a familiar type, activated by the cues present in the environment, and immediately identified a course of action that they knew from experience would work. They did not choose between options; they evaluated a single option by running a mental simulation of its consequences, and implemented it if the simulation suggested it would succeed.

Klein called this Recognition-Primed Decision making. The term basically refers to a rapid identification of a situation type based on experienced pattern recognition, followed by single-option evaluation rather than multi-option comparison. And he found it not just in firefighting but in military command, intensive care nursing, chess, and every other domain he studied where experts operated under genuine time pressure.

Gary Klein's research provides an essential counterbalance to the bias-and-heuristics tradition by demonstrating that rapid, intuitive decision-making is not inherently inferior to slow, deliberate analysis. Through the lens of Recognition-Primed Decision (RPD) making, Klein proves that in environments characterized by stable regularities and genuine time pressure, an expert's calibrated experience allows for superior accuracy.

Rather than weighing every possible option, the expert's brain utilizes sophisticated pattern matching to identify a viable course of action instantly, making intuition a refined tool of expertise rather than a mere cognitive shortcut.

The organizational implication drawn by Klein and endorsed by Susan is that decision-making architectures must be domain-sensitive to be effective. In high-velocity environments like clinical emergencies, attempting to force adherence to the full analytical ADEL framework is often counterproductive; instead, the Recognition-Primed Decision (RPD) making of experienced clinicians should be the primary trusted mechanism.

As a result, when errors occur in these contexts, the subsequent failure analysis should pivot away from *"analytical protocol compliance"* and focus instead on experience calibration and the refinement of feedback loops to sharpen future intuition.

Strategic planning sessions, capital allocation reviews, and competitive positioning decisions, where the domains are complex and novel, feedback loops are long, and the pattern recognition library of any single executive is inadequate, are where the full framework adds the most value. Explaining this to Jim and the others, Susan said:

"The question to ask before every major decision is not 'should I use my intuition or my analysis?' It is: 'Do I have extensive, well-calibrated experience in exactly this type of decision, and will I receive clear feedback on the outcome within weeks rather than months or years?' If the answer to both questions is yes, trust the Recognition-Primed system. If the answer to either question is no, use the framework."

Thinking About Thinking: The Metacognitive Imperative

The psychologist John Flavell coined the term "metacognition" in 1976 to describe thinking about one's own thinking, the capacity to observe, assess, and regulate one's own cognitive processes. Metacognition is what allows a student to recognize that she has not understood a passage she has just read, and to read it again. It is what allows a physician to notice that his diagnostic reasoning has stopped considering alternatives and to deliberately force himself back to a broader differential. It is what allows a CEO to recognize, in the middle of a committee deliberation, that the room is moving toward a conclusion faster than the evidence supports, and to slow it down.

Research consistently shows that metacognitive awareness is one of the strongest predictors of learning performance across domains. Learners who can accurately assess their own understanding learn more efficiently because they direct their effort toward genuine gaps rather than toward material they already know. Decision-makers who can accurately assess their own cognitive processes make better decisions because they can identify the moments when their System 1 is operating on insufficient evidence and engage System 2 before committing.

The moleskin practice that Susan establishes for Jim is, fundamentally, a metacognitive intervention. Writing down a decision before making it, articulating the alternatives, the evidence, the reasoning, the assumptions, is an act of observing one's own thinking from outside, and that observation is often sufficient to identify problems that the thinking itself would never have surfaced. The philosopher Blaise Pascal observed in the seventeenth century that "all of humanity's problems stem from man's inability to sit quietly in a room alone." Pascal was making a point about restlessness and distraction, but he was also describing something true about the benefits of deliberate, written reflection: it creates a distance between the decision-maker and the decision that makes honest assessment possible.

The Decision Record is institutionalized metacognition. It makes MegaHealth's collective thinking about its collective thinking into a systematic, retrievable, analyzable asset.

A Brief Interlude: The Philosopher's Gallery

Twenty-four centuries of thinking about thinking, what the great minds knew about decisions

This interlude steps outside the MegaHealth narrative to give the reader what Susan spent twenty years assembling: a working map of the intellectual tradition that makes Decision Sciences more than another management methodology. Each thinker below contributed something specific and practical to the discipline. None of them would have recognized the term *"Decision Sciences."* All of them were practicing it.

Socrates, 470–399 BC: The Man Who Knew He Knew Nothing

Socrates wrote nothing. Everything we know about him comes through the writings of others, primarily Plato. This is appropriate, because Socrates' central method was conversation, the dialectic, the back-and-forth of question and answer that he believed was the only genuine path to knowledge. The Socratic method doesn't give answers; it asks questions to test the validity of a claim. Through the disciplined process of uncovering assumptions, it moves us past the illusion of knowledge and into the genuine curiosity that marks the start of wisdom.

Socrates was executed in 399 BC by the city of Athens for the crime of corrupting the youth. His actual offense was something more specific: he had spent his life demonstrating, in public, that the most confident and powerful people in Athens, the politicians, the generals, the craftsmen who had built the Parthenon, did not actually understand the things they claimed to know.

He believed that they had opinions. They had techniques. They had success. But they did not have knowledge, justified, examined, coherent knowledge that could withstand the systematic questioning of the Socratic dialogue.

For Decision Sciences, the Socratic contribution is foundational: the principle that the first step in any genuine inquiry is the honest assessment of what you do not know. Every element of the ADEL framework's Assess phase is a Socratic exercise, the structured exposure of assumptions, the testing of claims against evidence, the systematic attempt to understand the limits of current knowledge before committing to action. The pre-mortem is Socratic dialogue accelerated and formalized: *"What do we not know about this decision that could cause it to fail?"*

The specific Socratic move that most directly maps onto Decision Sciences practice is what philosophers call the elenchus, the method of cross-examination in which Socrates accepts a claim for the sake of argument and then systematically extracts its implications until a contradiction emerges. The contrarian saint's advocate protocol in MegaHealth's structured deliberation is an organizational elenchus: it accepts the proposed decision for the sake of argument and then systematically constructs the strongest possible case that the decision is wrong, until the decision-making group has been forced to confront every significant vulnerability in the proposal.

Aristotle, 384–322 BC: The Practical Science of Good Judgment

If Plato was the philosophical father of Decision Sciences' epistemological dimension, the recognition that we operate on representations rather than reality, Aristotle was the father of its practical dimension. Aristotle's ambition was not just to understand how human beings think but to identify the conditions under which they think and act well, to develop what he called a practical science of human flourishing.

The concept that is most directly relevant to organizational Decision Sciences is phronesis, practical wisdom, or prudence. Phronesis is explicitly distinguished from both theoretical knowledge (episteme) and technical skill (techne). Take a look at these from the following perspectives:

- A doctor who knows all the biochemistry of a disease has episteme.

- A doctor who can perform all the procedures required to treat it has techne.

- A doctor who can look at a specific patient in a specific context with a specific history and make the right judgment about what to do has phronesis.

Phronesis is the capacity to deliberate well about the practical good under conditions of particularity and complexity that no general rule can fully anticipate.

Aristotle was clear that phronesis cannot be taught through books or lectures. It requires experience, specifically, the kind of experience that includes reflection, feedback, and the gradual refinement of judgment through repeated encounters with difficult cases. The implication for organizational Decision Sciences is direct.

What this means is that you cannot make an organization wise by teaching its members about decision theory. You make it wise by creating multiple aspects that include the:

- Conditions.

- Feedback loops.

- Psychological safety.

- Decision records.

- Retrospective protocols.

It's essential to know that all of these are aspects through which individual and collective judgment are continuously refined through experience.

The Decision Sciences system at MegaHealth is, in Aristotelian terms, a phronesis machine: an infrastructure for converting organizational experience into practical wisdom, for building the kind of deliberative capacity that cannot be purchased or copied but must be earned through disciplined engagement with reality.

Francis Bacon, 1561–1626: The Four Idols and the Science of Induction

Francis Bacon, Lord Chancellor of England and the philosopher most responsible for the scientific revolution's methodological foundation, devoted much of his philosophical energy to diagnosing the errors that prevent human beings from reasoning accurately about the world. In the Novum Organum (1620), he described four *"Idols"*, systematic distortions of thought that corrupt inquiry regardless of the intellectual quality of the inquirer.

The Idols of the Tribe are errors common to all human beings, the tendency to expect more order than exists, to perceive patterns in random data, to overweight supporting evidence and underweight contrary evidence. These are Kahneman's System 1 biases, described four hundred years before Kahneman.

The Idols of the Cave are individual idiosyncrasies, the specific biases and preconceptions of the individual's education, experience, and temperament. No two people have the same cave; the specific filters through which each individual processes information are unique, and the specific distortions they produce are correspondingly particular.

This is why the pre-deliberation independent estimation protocol, which captures each participant's unfiltered individual judgment before the group begins to converge, is so valuable: it produces evidence of the specific caves through which the room is filtering information.

The Idols of the Marketplace are errors introduced by language, the imprecision of words, the ambiguity of shared concepts, the ways in which the words we use to describe a problem shape the solutions we can imagine. The definition discipline that Susan brings to every decision, the insistence on precise language in the Decision Record, the explicit disambiguation of terms that different team members are using in different senses, is Bacon's remedy for the Idols of the Marketplace.

When it comes to healthcare, we must guard our strategy against *"Idols of the Theatre,"* the tendency to let ideological frameworks override empirical evidence. When we commit prematurely to a specific model of care or technological vision, we risk creating a strategic blind spot. Effective leadership requires us to dismantle these predetermined conclusions and remain open to evidence-driven transformation.

David Hume, 1711–1776: Causation, Correlation, and the Problem of Induction

David Hume, the Scottish Enlightenment philosopher, produced what many philosophers consider the most devastating critique of the basis of scientific reasoning ever formulated: the problem of induction.

Hume's argument was simple and devastating: our belief that the future will resemble the past, the assumption on which all learning from experience rests, cannot be justified by experience itself, because any attempt to justify it by experience would be circular. We believe the sun will rise tomorrow because it has always risen before. But the fact that it has always risen before is only evidence that it will rise tomorrow if we assume the principle of induction, that past regularities predict future ones, which is exactly what we are trying to justify.

For Decision Sciences practitioners, Hume's problem is not merely philosophical. It is eminently practical: every prediction we make from historical data rests on the assumption that the future will sufficiently resemble the past for the historical patterns to remain valid. This assumption is sometimes justified and sometimes not, and the skill of the decision analyst lies partly in knowing the difference.

The reference class forecasting technique that Susan uses, the practice of anchoring projections in historical base rates, is an explicit acknowledgment of Hume's insight: it considers the evidence that past cases provide while remaining alert to the possibility that the current case is sufficiently novel that the historical base rate may not apply.

The explicit statement of the reference class conditions, which makes the current case similar to the historical cases, is a direct engagement with the inductive validity question: *"Is the assumption that these cases are similar enough to support prediction actually justified?"*

Immanuel Kant, 1724–1804: The Categorical Imperative and Decision Governance

Immanuel Kant's moral philosophy, the most systematic and influential in the Western tradition, centers on what he called the Categorical Imperative. This principle suggests that before you act, you should ask yourself: "Would I be okay if everyone else did exactly what I am about to do?" If the answer is no, the action is immortal.

Kant's point was not merely utilitarian, not simply that we should consider the consequences of our actions for others, but that the form of moral action requires a kind of universalizability: the willingness to accept that everyone in relevantly similar circumstances should be bound by the same principles that govern your action.

At its core, the governance of Decision Sciences is a Kantian project. It seeks to establish principles that are truly universalizable, meaning that the rules that maintain their integrity across diverse cases, actors, and eras. By prioritizing consistency over expediency, this framework guards against 'governance by convenience,' where principles are championed only when they serve immediate interests and discarded the moment they become an obstacle.

The Categorical Imperative has a specific and practical application to the Minority Report problem that MegaHealth encountered. The question Susan posed, *"what principles should govern our use of predictive analytics in personnel decisions?"*, is a Kantian question: not *"what would produce the best outcome in this specific case?"* but *"what principle, applied universally to all cases of this type, would we be willing to endorse?"*

The answer that the team reached, *"predictive analytics inform but never constitute the sole basis for decisions that affect individuals,"* is a Kantian maxim: a principle that can be applied universally, consistently, and transparently, and that all affected parties could reasonably accept as the governing rule.

Charles Sanders Peirce, 1839–1914: Pragmatism and the Test of Consequences

The American philosopher Charles Sanders Peirce, founder of the philosophical school of pragmatism, proposed what he called the pragmatic maxim: *"Consider what effects, that might conceivably have practical bearings, we conceive the object of our conception to have. Then, our conception of those effects is the whole of our conception of the object."* In plain terms: the meaning of a concept is fully captured by its practical consequences. Ideas that make no difference in practice have no genuine cognitive content.

Applied to Decision Sciences, Peirce's maxim is the intellectual foundation of outcome-based learning: the only way to test whether a decision was made well is to examine its consequences, not the process by which it was made, but the actual effects it produced in the world. Process matters, for Peirce's descendant William James, because it is the reliable predictor of consequences, but the ultimate test is always practical.

Susan's insistence on pre-specifying outcome metrics is a masterclass in pragmatism. It shifts the post-mortem from an abstract debate over process to a concrete assessment of impact. By moving the focus from the decision as an isolated act to its consequences as empirical evidence, she ensures that organizational learning is a cumulative, data-driven evolution rather than a series of retrospective rationalizations.

Ludwig Wittgenstein, 1889–1951: The Limits of Language and the Power of Practice

Ludwig Wittgenstein, arguably the most important philosopher of the twentieth century, is not usually cited in business books. His work is dense, often cryptic, and frequently misappropriated in ways that he would have found appalling. But one of his central insights, expressed most accessibly in his later work Philosophical Investigations, is directly and practically relevant to organizational decision-making.

Wittgenstein argued against the idea that meaning is a private mental state that language merely expresses. Meaning, for Wittgenstein, is a matter of use, of the practical role a word or concept plays in the form of life that gives it significance. The implications are radical: you cannot fully understand a concept by reading a definition of it. You understand it by using it, by participating in the practices that give the concept its significance.

For Decision Sciences, this means that the concepts in the framework, decision quality, cognitive bias, option value, reference class, are not fully understood until they are used in practice. Reading about anchoring does not calibrate your decision-making against anchoring. Making decisions, writing down your reasoning, receiving feedback on your predictions, and adjusting your process based on that feedback, is what calibration actually is. The Decision Record is not an educational tool. It is a practice. And practice, in Wittgenstein's sense, is the only genuine teacher.

This is why Susan does not begin Jim's Decision Sciences education with theory. She begins it with a coin flip. The coin flip is a practice, a direct, experiential confrontation with the gap between felt confidence and actual accuracy that no amount of reading about overconfidence could produce. The moleskin is a practice. The pre-mortem is a practice. The calibration exercise is a practice. The framework is learned by doing it, not by learning about it. Wittgenstein would have recognized and approved.

Thomas Kuhn, 1922–1996: Paradigm Shifts and Organizational Epistemology

Thomas Kuhn's 1962 book The Structure of Scientific Revolutions introduced to the intellectual mainstream a concept that has become one of the most widely used, and most frequently misused, in modern discourse: the paradigm shift. Kuhn's argument was that normal science proceeds within a paradigm, a framework of assumptions, methods, and exemplary problems that defines what counts as a valid question and a valid answer in a given scientific community. Paradigm shifts occur when anomalies accumulate to the point where the existing paradigm can no longer accommodate them, and a new paradigm replaces it through a process that is at least partly sociological rather than purely rational.

The organizational Decision Sciences application of Kuhn's insight is that every organization operates within a strategic paradigm, a set of assumptions about its market, capabilities, competitors, and value proposition that shapes what it notices, what it considers relevant, and what decisions it is capable of making.

The paradigm is not wrong, exactly; it captured genuine patterns that were valid at some point. But as the environment changes, the paradigm's anomalies accumulate, and the organization that cannot shift its paradigm will continue making decisions that are locally coherent but globally mistaken.

The MegaHealth decision database functions as a paradigm anomaly detector. While a single retrospective might excuse a poor result as an outlier, the systematic comparison of thousands of outcomes reveals deeper patterns of error. When the organization is consistently wrong in the same direction, it is no longer a matter of bad luck; it is a signal that the strategic paradigm itself is failing and requires fundamental revision.

CHAPTER THREE: THE FRAMEWORK OF DECISION MAKING

How Aristotle's phronesis becomes operational infrastructure

The Stoics and the Architecture of Deliberation

The Stoic philosophers of ancient Greece and Rome did something that most of their philosophical contemporaries did not: they treated virtue not as an innate quality of soul but as a practice, a set of disciplines that had to be actively cultivated, daily, through structured reflection and deliberate attention. For Epictetus, the former slave who became one of the most influential philosophers of the Roman era, the fundamental philosophical task was not to understand the world but to understand what was within your power to control and what was not, and to align your responses accordingly.

For Marcus Aurelius, the philosopher-emperor who governed the Roman Empire during one of its most turbulent periods, the Stoic practice took the form of a daily journal, the Meditations, written for his own use rather than for posterity, in which he examined his own reasoning, held his decisions up to scrutiny, and applied the principles of Stoic philosophy to the specific, gritty challenges of governing an empire while fighting wars on multiple fronts.

"You have power over your mind, not outside events. Realize this, and you will find strength."

-Marcus Aurelius, Meditations

What Marcus Aurelius was practicing, in the privacy of his tent on the Danube frontier, was the earliest known version of what Decision Sciences calls the decision record: a systematic external inscription of internal deliberation that creates accountability, enables reflection, and builds the habit of distinguishing what one can control from what one cannot. The moleskin that Susan gives Jim is a modern version of the same tool.

The Stoic philosophers understood that good decision-making requires regular practice, external discipline, and genuine humility about the limits of one's own reason. Two thousand years of subsequent research has confirmed that they were right.

Herbert Simon and the Architecture of Organizations

The twentieth century produced its own version of the Stoic insight in the work of Herbert Simon, the polymath who won the Nobel Prize in Economics in 1978 and whose career spanned economics, psychology, cognitive science, computer science, and political science. Simon's most influential concept, bounded rationality, described the fundamental constraint on human decision-making: we do not optimize; we satisfice. Faced with a complex decision, we do not evaluate all available options and choose the one with the highest expected utility. We search through options until we find one that meets our aspiration level, one that is good enough, and then we stop searching.

This observation has profound organizational implications. If decision-makers focus on satisfaction rather than optimization, then the quality of their decisions depends critically on the structure of their search process, on how they define the aspiration level, how broadly they search, what order they search in, and when they stop. A process that seeks satisfaction combined with a well-designed search structure can produce near-optimal outcomes. A poorly structured satisfying process creates a 'rationality illusion.' By limiting search to familiar territory or deferring to authority signals, the system reaches a local optimum while ignoring superior alternatives. This creates a cognitive paradox: the decision-maker feels they have been rigorous, even as the structural flaws in their search process guarantee a sub-optimal result.

Simon's work also pioneered the study of organizational decision-making as distinct from individual decision-making. Organizations do not decide the way individuals do. They have standard operating procedures, authority hierarchies, information routing systems, and cultural norms that shape every decision made within them, often more powerfully than the individual cognition of the people nominally making the decisions. Building a Decision Sciences framework for an organization requires understanding and deliberately redesigning these organizational decision processes, not just educating individual decision-makers about cognitive bias.

Susan's Framework: The Second Week

When Susan arrives the following Tuesday, she carries with her a single printed page and a laptop. She does not carry a PowerPoint deck with sixty slides. She has learned, over years of working with executives, that sixty slides produce sixty minutes of slide-reading and almost no genuine thinking. One page produces conversation and on this page she has a simple diagram: five boxes connected by arrows. These boxes, labeled from left to right, include: :

- DATA

- ANALYTICS

- INSIGHTS

- DECISIONS

- LEARNING

The arrows between them are bidirectional. Below the diagram, a single annotation that reads:

"The decision is not the end. It is the middle."

Susan, *"I want to start with what we already have, because one of the things I have noticed about most executives is that they underestimate how much of the hard work is already done. You have built the data layer with Omnimics. You have built the analytics and insight layer with Vairos. The infrastructure is there.*

What we are adding today is the decision layer, and the learning layer that flows from it."

Jim, *"And we don't have those yet."*

Susan, *"We have fragments. We have post-mortems that happen sometimes. We have executive retrospectives that happen occasionally. We have a culture in which people occasionally, and informally, connect past decisions to present outcomes. What we do not have is a system. A disciplined, persistent, organization-wide infrastructure that makes decision tracking as routine as revenue tracking."*

Jim, *"Is there any organization in healthcare that has this?"*

Susan, *"Fragments of it, in the most advanced health systems. Kaiser Permanente has elements of it in their clinical decision support infrastructure. Geisinger has pieces of it in their medical management programs. But a fully integrated decision intelligence layer, connecting operational, strategic, and clinical decisions in a single feedback system, I don't know of anyone who has built it completely. Which means that if we build it here, we will have a genuine first-mover advantage. Not in a temporary, quickly-replicated way. In a structural, compound-over-time way, because our decision database will get smarter every quarter, and competitors who haven't started can't catch up quickly."*

The Five Intellectual Foundations of Decision Sciences

Before Susan presents the operational framework, she spends time ensuring that Jim understands the intellectual geography of Decision Sciences. She wants him to know where the discipline comes from, and why its multidisciplinary character is a feature rather than a limitation.

Foundation 1: Decision Analysis and Operations Research

Decision analysis emerged as a formal discipline at Harvard and MIT in the 1960s, largely through the work of Howard Raiffa and Robert Schlaifer. It provided the mathematical backbone for structured decision-making in uncertainty. This included things like decision trees, influence diagrams, expected utility calculations, and the formal treatment of probability and risk.

Operations research, a related discipline developed during World War II to optimize military logistics and expanded postwar to commercial and industrial applications, contributed linear programming, queuing theory, simulation, and optimization algorithms to the toolkit. Together, these disciplines provided the quantitative infrastructure for systematic decision analysis.

The critical limitation of classical decision analysis is its assumption that the relevant options, outcomes, and probabilities can be adequately specified in advance. However, this is an assumption that holds well in some domains and breaks down completely in others. Real strategic decisions often involve options that are not known in advance, outcomes that cannot be adequately predicted, and probabilities that resist reliable estimation. These are all limitations addressed in the foundations of decision sciences mentioned below. .

Foundation 2: Behavioral Decision Theory

As documented in Chapter Two, behavioral decision theory, developed primarily by Kahneman, Tversky, Thaler, and their colleagues, demonstrates that actual human decision-making departs systematically from the prescriptions of classical decision analysis. Its contribution to Decision Sciences is diagnostic. What this means is that it explains why decision processes fail and provides the theoretical basis for designing interventions that compensate for predictable cognitive errors.

Foundation 3: Causal Inference and Econometrics

Understanding whether a decision actually caused the outcome observed, as opposed to merely correlating with it, requires the tools of causal inference. This is one of the most technically demanding and practically important dimensions of Decision Sciences, and it is one that most organizations haven't started working on yet..

The causal inference problem is fundamental. What happens here is that when you decide and observe an outcome, how do you know whether the decision caused the outcome? Maybe the outcome would have occurred anyway, without your decision. Maybe another factor, a change in the market, a regulatory shift, a competitor's move, caused the outcome and your decision was irrelevant. Without a valid causal inference method, your organizational learning is building on a foundation of coincidental patterns rather than actual cause and effect.

The major tools of causal inference, randomized controlled trials, difference-in-differences analysis, instrumental variables, regression discontinuity designs, and propensity score matching, are well-established in academic research and increasingly available to organizational practitioners through modern data science platforms.

As we continue, we'll learn that Susan's decision tracking infrastructure will incorporate basic causal inference designs where feasible and will at a minimum flag the causal ambiguity in decisions where clean causal identification is not achievable.

Foundation 4: Neuroscience and Cognitive Science

Neuroscience research, using imaging, biological studies, and computational models, now reveals the biological reasons why human judgment succeeds or fails. Researchers have mapped the specific brain circuits that manage how we calculate value, assess risk, and process social information. This work shows how our brains integrate emotional signals with analytical thinking to reach a final decision.

For organizational practitioners, the most important neuroscientific finding is the confirmation of the dual-process architecture: two systems, with different speeds, different accuracy profiles, and different conditions of reliability, constantly competing to govern behavior. Building decision architectures that engage the right system for the right decision type is the core practical implication of this neuroscientific foundation.

Foundation 5: Organizational Behavior and Group Dynamics

The fifth and final foundation of Decision Sciences is organizational behavior, the study of how groups make decisions, how organizational structures and cultures influence decision quality, and what interventions can improve collective judgment.

This research covers several key concepts in team dynamics: groupthink, psychological safety, and team performance. It also explores information cascades, which occur when group discussions amplify individual errors because members simply follow the opinions of those who spoke before them. In short, we now have a deep understanding of why groups either excel or fail based on how they process information and interact with each other.

One finding from this literature is particularly important for healthcare leadership is the relationship between team diversity and decision quality. Based on the finding, teams with diverse cognitive styles, domain expertise, and demographic backgrounds make better decisions on complex tasks than homogeneous teams, not despite the difficulty of reconciling diverse perspectives, but because of it. This research shows that the tension created by a truly diverse team is not a sign of failure. Instead, it serves as a necessary mechanism that forces deeper analysis. By bringing different perspectives to the table, we uncover critical information that uniform teams would likely overlook.

The ADEL Framework: Assess, Decide, Execute, Learn

After having established the intellectual foundation, Susan presented the operational framework that will structure MegaHealth's Decision Sciences implementation. She calls it ADEL, four phases that together span the full lifecycle of a significant organizational decision.

Phase	Core Question	Key Tools	Output
Assess	What do we actually know, and what are we uncertain about?	Reference class forecasting, stakeholder mapping, pre-mortem analysis, competitive scenario planning	Decision Brief: structured evidence summary
Decide	What should we commit to, and on what basis?	Decision Record, weighted scoring, devil's advocacy, structured deliberation	Formal Decision Record with rationale and assumptions

| Execute | What downstream decisions and actions does this commitment trigger? | Decision Cascade mapping, RACI matrix, milestone governance | Decision Cascade with 47+ enabling decisions mapped and owned |
| Learn/ Track | Did the decision produce the expected outcome, and what do we learn from the difference? | Outcome tracking dashboard, retrospective analysis, causal assessment | Learning Record fed back into future assessments |

Phase 1: Assess, The Art of Knowing What You Know

The Assess phase is the pre-decision intelligence phase. It is where the evidence for a decision is systematically assembled, the cognitive biases most likely to distort the decision are identified and countered, and the range of genuine alternatives is defined.

Most organizations do some version of this, they commission analyses, conduct market research, and review financial models. However, what they rarely do is design the assessment process to counteract the specific biases most likely to be active in that specific decision context. Susan's contribution is to make this bias-counteraction explicit and systematic.

The most important bias-counteraction technique in the Assess phase is the pre-mortem, a structured exercise developed by psychologist Gary Klein, in which a decision team is asked to imagine that a proposed decision has been implemented and has failed spectacularly, and then to work backward to identify the causes of failure. The pre-mortem is counterintuitive. It asks people to imagine failure before the decision is made, which feels defeatist and contrary to the optimistic culture of most leadership teams. But its effect on decision quality is significant. By discussing potential failure scenarios before committing to a plan, we can uncover hidden risks and assumptions. This process, known as a pre-mortem, helps us counteract the natural overconfidence and optimism that often cloud our judgment. It ensures we address critical flaws before they become real problems.

Reference class forecasting, the practice of anchoring projections in the actual historical base rates of similar decisions, is the second critical tool of the Assess phase. What this means is that for every major decision at MegaHealth, Susan's protocol requires the identification of a reference class: a set of past decisions similar in kind to the current one, whose actual outcomes can be analyzed to calibrate the current projection. So, in practicality, if the base rate of cardiology expansion projects achieving their Year 3 revenue projections is 40 percent, this fact must be explicitly incorporated into the current projection, it may not simply be overridden by the optimism of the project team.

Phase 2: Decide, The Discipline of Commitment

The Decide phase is the formal choice point. It is where the evidence assembled in the Assess phase is converted into a commitment, where the analysis becomes accountable.

The central tool of the Decide phase is the Decision Record, a structured document that captures the full context of a decision in a form that enables future review and learning. Given this, Susan designed MegaHealth's Decision Record template with twelve required fields that included:

- Decision ID
- Decision Title
- Decision Owner
- Decision Date
- Decision Type
- Urgency Level
- Decision Statement
- Alternatives Considered
- Selection Rationale
- Key Assumptions
- Success Metrics

- Review Triggers

Each of these fields does specific cognitive work. For example:

- The alternatives considered field forces the decision-maker to articulate what was not chosen and why, countering the confirmation bias tendency to focus exclusively on the chosen option's merits.

- The key assumptions field makes explicit the beliefs on which the decision rests, enabling future review to assess whether the assumptions were correct.

- The success metrics field creates pre-committed criteria for evaluating the decision's outcome, preventing the post-hoc rationalization of poor results.

- The review triggers field is perhaps the most innovative as it pre-specifies the conditions under which the decision should be revisited, market changes, performance thresholds, time milestones, or specific events that would warrant reassessment. This field directly counteracts the sunk cost fallacy by establishing, before any investment is made, the criteria for abandonment.

While explaining all this Susan, concluded with:

"The Decision Record is not paperwork. It is a commitment device, a deliberate constraint on the future self that overrides the sunk cost bias and the escalation of commitment. By specifying in advance the conditions under which you will reconsider, you make it harder for your future self to rationalize continued investment in a failing decision. You are binding Ulysses to the mast before you hear the Sirens, not because you distrust your future judgment, but because you understand precisely how it will be compromised."

The Structured Deliberation Protocol

Beyond the Decision Record, the Decide phase uses a structured protocol to guide how groups make choices. This process is built on organizational behavior research and includes four specific interventions designed to stop common group biases. By following this protocol, we can ensure that team discussions remain objective and focused on the best possible outcome. These interventions include:

1. **Pre-deliberation independent estimation:** Before any group discussion, each participant is asked to write down their individual assessment of the key decision variables, the most likely outcome, the probability of success, the most important risk, without knowledge of others' assessments. These estimates are collected and aggregated before deliberation begins. This technique directly counters information cascades and authority bias: it ensures that the group discussion starts from the distribution of genuine independent judgments rather than from the anchor of whoever speaks first (typically the highest-ranking person in the room).

2. **Structured contrarian advocacy (or devil's advocate):** One member of the decision team is assigned the contrarian advocate role, explicitly tasked with constructing the strongest possible case against the emerging consensus option. The contrarian's advocate role is rotated to prevent it from becoming associated with a specific individual (which would reduce its effectiveness), and its contributions are explicitly valued rather than merely tolerated. Research by Charlan Nemeth at Berkeley demonstrates that minority dissent, even when it is wrong, improves group decision quality by disrupting premature closure and stimulating more thorough consideration of the option space.

3. **Dialectical inquiry:** For particularly high-stakes decisions, Susan uses a technique called dialectical inquiry, a structured debate format in which two sub-teams develop and present opposing recommendations, each building the strongest possible case for their position. The group then synthesizes the best arguments from both positions. This is a more intensive version of devil's advocacy, drawn from a philosophical tradition that goes back at least to Hegel's dialectic and more practically to the adversarial inquiry processes used in judicial and military decision-making.

4. **Explicit assumption surfacing:** Before the decision is finalized, the group is walked through the key assumptions underlying the recommended option and asked, "If this assumption turns out to be

wrong, does that change our decision?" This simple question surfaces hidden assumptions, tests the robustness of the recommendation to assumption violations, and identifies the most important uncertainties for future monitoring.

Phase 3: Execute, The Decision Cascade

The Execute phase is where most decision frameworks end, and where most organizational value is lost. The strategic decision is announced; the enabling work begins; and, almost immediately, the connection between the strategic rationale and the operational choices required to implement it begins to fray.

Susan's Execute phase is built around a single, powerful concept known as the decision cascade. Every significant strategic decision trigger a cascade of enabling decisions, each one a real commitment with its own stakeholders, timeline, dependencies, and success criteria. Most organizations manage these enabling decisions informally, treating them as implementation details. Susan, however, treats them as decisions, with all the governance, tracking, and accountability that the ADEL framework applies to the strategic decision itself.

The decision cascade for MegaHealth's CardioNorth Initiative, the expansion of cardiology services into the northeast service area, contains forty-seven distinct enabling decisions across eight workstreams. Each of these decisions is logged in the decision tracking system with its own record that includes information about the owner, timeline, alternatives considered, assumptions, and success criteria. This is not administrative overhead. It is the mechanism by which a strategic vision becomes operational reality rather than an announcement in a board presentation.

Explaining this Susan said, *"Jim, here is what I want you to understand about the Execute phase. When you decided to pursue CardioNorth, you did not make one decision. You triggered forty-seven decisions. Real estate, staffing, clinical model design, technology, marketing, financial structure, legal compliance, and operations. Each of these decisions is a commitment of resources. Each has a decision owner, a timeline, alternatives, assumptions, and expected outcomes. And each of them can succeed or fail independently. If fifteen of your forty-seven decisions are made well and thirty-two are made poorly, CardioNorth will fail, not because the strategy was wrong but because the decision cascade was unmanaged."*

Jim, *"We have never mapped decisions this way before."*

Susan, *"Almost no organization does. And that is why most strategic initiatives fail at the execution stage. It is not a strategy failure. It is decision cascade failure, the compounding of dozens of small, untracked, unaccountable enabling decisions into an outcome that looks, retrospectively, like a strategy that didn't work."*

Phase 4: Learn, Where Intelligence Becomes Wisdom

The Learn phase closes the loop since it is the mechanism by which decisions and their consequences become organizational intelligence, by which the wisdom of the organization compounds over time rather than evaporating with each leadership transition or strategic pivot.

Learning from decisions requires three things that most organizations do not systematically provide. These include:

- It requires accurate outcome measurement, the actual documentation of what happened, in the specific, quantitative terms that the original Decision Record specified.

- It requires honest comparison of outcomes to predictions, not the retrospective rationalization of discrepancies but the frank acknowledgment of them and the disciplined inquiry into their causes.

- It requires institutional memory, the preservation and accessibility of Decision Records across leadership transitions, so that the reasoning of previous decision-makers is available to inform current ones.

In 1910, philosopher John Dewey described learning as a continuous cycle of experience, reflection, conceptualization, and experimentation. While Dewey focused on individuals, this same cycle applies to how organizations learn.

In this model, an organization executes a plan and sees the results. Our team then reflects on the difference between what we expected and what happened. From there, we develop theories to explain that gap and experiment with new processes in the next cycle. The ADEL framework takes this philosophy and turns it into a practical system for organizational decision-making.

The Linac Decision: Framework in Action

Jim's most pressing decision, the radiation therapy equipment procurement, arrives three weeks after Susan's framework presentation, and it arrives with the specific complexity that tests every element of the ADEL process.

What's happening now is that MegaHealth operates four linear accelerators, the machines that deliver precision radiation therapy for cancer patients. The machines are thirteen years old. A research paper that Kelsey Henderson, Director of Radiation Oncology, has brought to Jim's attention reports that radiation toxicity risk begins to increase significantly in machines older than fourteen years, as component degradation affects beam accuracy.

Two vendors have been invited to present. Susan insists that before the vendor presentations, she and Jim work through the Assess phase.

The Assess Phase: Before the Vendors Arrive

Susan runs the pre-mortem.

Susan, *"Imagine it is three years from now. We made a decision about the radiation equipment. It went badly. Write down the three most likely reasons it went badly."*

Jim writes:

1. We overspent on equipment that turned out not to differentiate us clinically.

2. The implementation disrupted operations and we lost patient volume during the transition.

3. The referring oncologists didn't adopt the new capabilities.

Susan, *"Now: imagine the opposite. Three years from now, the decision went spectacularly well. What happened?"*

Jim writes:

1. The new capabilities attracted high-complexity cases we currently refer out to our competitors.

2. We established a reputation for radiation oncology excellence that differentiates us.

3. The co-marketing from the vendor brought patient volume we couldn't have generated independently.

Susan, *"Now look at both lists. What are the assumptions embedded in them?"*

This question produces twenty minutes of intensive thinking. Jim identifies seven core assumptions underlying any favorable outcome. These assumptions include:

1. Financing terms are sustainable

2. Implementation timeline is achievable

3. Co-marketing support is genuinely valuable

4. Patient population in the service area can support the volume projections

5. New capabilities genuinely outperform the current ones in clinical outcomes

6. Referring oncologists are dissatisfied enough with current capabilities to change their behavior

7. No competitor enters with superior equipment in the eighteen-month window between decision and full operation.

Susan, *"Seven assumptions. Any one of them, if wrong, materially changes the decision. How many of them have we actually verified with evidence, versus assumed based on intuition?"*

Jim counts. Three have been verified with some rigor. Four are, as he puts it, *"hopes."*

Susan, *"Before the vendors come in, I want to close two of those gaps. I want to talk to three referring oncologists about their equipment preferences, and I want to pull the volume data on complex radiation cases that we currently refer out of our system. Two conversations and two data pulls. That is a week's work. Do we have a week?"*

Jim, *"We have a week."*

What the Referring Oncologists Said

The three conversations with referring oncologists produce a finding that surprises both Jim and Kelsey. Two of the three oncologists are indeed dissatisfied with MegaHealth's current equipment, but their dissatisfaction is not primarily about the equipment age. It is about the treatment planning software, which they describe as slower and less intuitive than competitors. The equipment itself, they suggest, performs adequately for most of the cases they refer to.

The volume data tells an equally nuanced story. Of the complex radiation cases MegaHealth refers to, approximately 60 percent go to competitors with newer equipment, but the primary driver of referral is not equipment but geography: the patients live closer to the competitor's facility.

Susan, *"The assumption that new equipment will recapture referred volume turns out to be approximately 40 percent correct and 60 percent wrong, based on what we just learned. The equipment matters, but geography matters more, and we cannot change geography with a capital purchase. Does this change the decision?"*

Jim, *"It changes the financial case. Our base-case revenue projection assumed we would recapture about 30 percent of referred volume. If geography is the primary driver of referral, we should probably assume 12 to 15 percent recapture."*

Susan, *"Which changes the NPV by how much?"*

Jim, *"Significantly. We need to rerun the model."*

Susan, *"This is what the Assess phase does. It does not guarantee a good outcome. It reduces the probability of a bad one by converting assumptions into evidence. The vendors are coming tomorrow. We will meet them with a fundamentally more accurate model than we would have had without this week's work."*

The Vendor Presentations and the Kelsey Moment

The two vendor presentations take place in the MegaHealth boardroom, two hours apart, on a Thursday morning. Susan has arranged the room deliberately: the financial analysis is face-down on the table, to be revealed only after the clinical and operational presentations.

This is a de-anchoring measure; she wants the committee to form their clinical and operational assessments before being exposed to the price points.

Vendor A presents first and opens with a comprehensive fleet replacement proposal: all four linear accelerators replaced simultaneously, with vault upgrades, for $50 million. The clinical case is compelling, a quantum improvement in beam accuracy, the latest adaptive radiation therapy capabilities, and a vendor-guaranteed commissioning timeline. The presentation is polished, confident, and expensive.

Vendor B presents a phased approach: replace the two oldest machines in Year 1, with the newest adaptive therapy capability on one of them, for approximately $12 million upfront and $6 million per machine over the following eight years. The clinical case is more modest but credible: immediate risk mitigation on the oldest equipment, one cutting-edge machine to support complex cases and market differentiation, and a financing structure that preserves capital for other priorities.

Jim invites Kelsey to join him and Susan in the boardroom after both presentations for the decision session.

What follows is one of those moments that Susan will later describe as the clearest illustration she has ever seen of the behavioral economics of organizational decision-making in real time.

Jim, *"Kelsey, if budget were not a constraint at all, what would you recommend?"*

Kelsey Henderson, Director of Radiation Oncology, *"Option A. Full replacement. It puts us in a clinical leadership position immediately and gives us capabilities that none of our regional competitors have. I could recruit on the back of that equipment."*

Jim, *"Thank you. Now, if I told you that Option A will require a ten percent reduction in capital spending in other clinical departments this year, what would you recommend?"*

Kelsey hesitates. The hesitation is brief but visible. Susan is writing.

Kelsey Henderson, *"In that case... I would probably choose Option B. The new adaptive therapy machine gives us a meaningful differentiator, the financial pressure on other departments is a real concern, and the phased approach lets us assess the clinical benefits before committing to the full fleet."*

Susan, *"Kelsey, thank you."*

What just happened in this conversation is one of the most important things in all of Decision Sciences. You gave two different answers to the same objective question, because the framing changed. Neither answer is dishonest. Both answers are completely rational.

But they are rational in different reference frames. The first answer is rational in the reference frame of pure clinical optimization. The second is rational in the reference frame of clinical optimization with organizational resource constraints.

"The question that Decision Sciences asks is: which reference frame is the right one? And the answer, almost always, is that the real decision is the second one, because real decisions always have real constraints, and the job of a decision framework is to make those constraints explicit from the beginning, not to introduce them as correctives at the end, says Susan.

Kelsey is silent for a moment.

Kelsey Henderson, *"I have never had a conversation like this at any health system in twenty years of practice."*

Susan, *"Most organizations don't have this conversation. They have the first version, the clean, budget-unconstrained version, and then they have a separate budget conversation that overrides the clinical recommendation, and nobody explicitly reconciles them. The decision that gets made is neither the clinically optimal nor the financially optimal. It is whatever emerges from the friction between two separate conversations that were never designed to connect. That is not a decision process. It is organized confusion."*

The Decision Record

Jim selects Option B. Susan writes it in the Decision Record before he leaves the room.

Decision ID	MH-CAP-2025-0014
Decision	Replace two oldest linear accelerators in Year 1 (Option B phased procurement). Full fleet replacement deferred to 5-year capital plan.
Owner	Jim, CEO
Key Assumptions	1) New adaptive therapy capability attracts 12-15% of currently referred complex cases. 2) Phased implementation does not disrupt current patient volumes. 3) Vendor co-marketing support generates 200+ qualified patient inquiries in Year 1. 4) Year 2-4 replacements are financially viable at current capital allocation rates.
Success Metrics	<ul><li>Volume: 8% increase in radiation oncology cases by Month 18.</li><li>Revenue: $400K incremental revenue by end of Year 1.</li><li>Quality: Zero radiation incidents attributable to equipment.</li><li>Recruitment: One additional radiation oncologist hired within 12 months of installation.</li></ul>
Review Triggers	If Year 1 volume increase is below 4% by Month 12, convene a full clinical and competitive review before proceeding with Year 2 machine replacements.

Jim *"That is the most complete record I have ever made of a capital decision."*

Susan, *"In eighteen months, when we make the next machine replacement decision, you will have actual data against every one of those success metrics.*

You will know whether the assumption about referred volume recapture was right. You will know whether the co-marketing worked. You will know whether the clinical recruiter can use the equipment as a differentiator. And you will make a better decision as a result, not because you are smarter, but because you are more informed. That is what Decision Sciences does. It makes organizations smarter over time by connecting their decisions to their evidence."

On the Nature of Frameworks: What Confucius and Kant Understood

Before we leave the framework chapter, it is worth spending a moment on a question that practical-minded readers may be asking: is all this framework necessary?

Can't experienced leaders simply make good decisions without a systematic process?

The answer is nuanced. In some domains, yes: genuine expertise produces reliable intuitions that outperform explicit deliberation. The experienced chess grandmaster, the expert diagnostician, the seasoned trauma surgeon operating in familiar territory, all produce better outcomes from their trained System 1 than they would if they stopped to deliberate through every move.

But the domains in which organizational executives typically operate, strategic planning, capital allocation, market expansion, talent management, organizational design, are domains of genuine complexity and novelty. These are domains where the feedback loops are long, the signals are noisy, and the environmental regularities that calibrate intuition are absent. These are exactly the domains where the behavioral economics research shows the most reliable failures of unstructured intuition, and where systematic frameworks produce the most reliable improvements.

Confucius, in the Analects, described the progression of wisdom in a way that maps almost perfectly onto the relationship between framework and expertise:

> *"At fifteen, I had my mind bent on learning. At thirty, I stood firm. At forty, I had no doubts. At fifty, I knew the decrees of Heaven. At sixty, my ear was an obedient organ for the reception of truth. At seventy, I*

could follow what my heart desired, without transgressing what was right."

The framework is for the fifteen-year-old. The spontaneous rightness of the seventy-year-old comes after decades of disciplined practice, honest feedback, and accumulated wisdom, not despite the structure but through it. You cannot skip to seventy. You must earn it.

In the Groundwork for the Metaphysics of Morals, Immanuel Kant argued that acting from a core principle, rather than just following a feeling, is the true expression of a good will.

A person who does the right thing only when they feel like it is not consistently dependable. In contrast, someone who internalizes principles to guide their actions, regardless of their mood, is truly reliable. Our decision framework acts as the practical version of Kant's moral theory: it is a structured commitment that ensures high-quality judgment, even when our intuition is absent or clouded.

"I do not claim that frameworks are superior to wisdom. I claim that frameworks are the path to wisdom, for organizations that lack the centuries of accumulated experience that produce reliable institutional intuition, and for decisions that occur in domains where even the most experienced leaders cannot reliably distinguish genuine insight from sophisticated rationalization," said Susan.

The Pre-Mortem in Practice: Thirty Minutes That Save Millions

Gary Klein developed the pre-mortem technique in the 1990s as a direct response to the well-documented failure of conventional risk analysis. Conventional risk analysis, in which a team is asked to identify risks before a project begins, produces systematically incomplete lists, for two reasons that include:

1. The people doing the analysis are already committed, at some level, to the project they are analyzing, and their commitment suppresses their ability to imagine failure scenarios.

2. The social dynamics of risk brainstorming in groups tends toward anchoring on the first risks mentioned, underweighting low-probability/high-impact scenarios, and self-censoring on risks that challenge the project champion's central assumptions.

The pre-mortem solves both problems with a simple reframing. Instead of asking *"what could go wrong?"*, a question that activates risk-identification against the background assumption of success, it asks *"imagine it has gone wrong, spectacularly. It is eighteen months from now and this project has failed. What happened?"* This reframing does three things. It normalizes failure as a possibility, reducing the social cost of pessimistic contributions. It activates a different kind of pattern recognition, not the pattern recognition of risk analysis, which identifies abstract categories of risk, but the pattern recognition of narrative imagination, which generates specific, concrete, vivid failure scenarios.

And it forces the identification of causal chains, not just *"the technology failed"* but *"the technology failed because we underestimated the integration complexity, which we should have known because our last three EHR integration projects all took twice as long as projected."*

The practical value of the pre-mortem is its ability to surface information that participants hold but are not contributing. In research by Deborah Mitchell and colleagues at the Wharton School, pre-mortem groups generated 30 percent more unique risks than conventional risk analysis groups working with the same information. The increase in identified risks did not come from the same people thinking of more problems. Instead, it came from participants who previously kept their concerns private. While these individuals were reluctant to speak up during a standard risk analysis, they felt comfortable contributing during a pre-mortem, where the process explicitly encourages and rewards a skeptical perspective.

Susan runs a pre-mortem for every decision that goes through the full ADEL framework. The session is thirty minutes. It is always the first thing, before any other decision analysis, and it reliably generates at least one significant insight that was not in any prior analysis, one risk, assumption violation, or failure mode that would not have been surfaced without the explicit license to imagine failure.

Benjamin Franklin's Ledger: The Oldest Decision Framework in American History

In 1772, the British scientist and political philosopher Joseph Priestley, later famous for his discovery of oxygen, wrote to his friend Benjamin Franklin asking for advice on a difficult personal decision he was facing. Franklin's reply has become one of the most quoted documents in the history of practical decision-making.

Franklin described his practice of *"moral algebra,"* dividing a sheet of paper into two columns, Pro and Con, and then listing, over the course of several days, all the arguments he could identify on each side. After listing them, he would go through the list and *"endeavor to estimate their respective Weights,"* not by numbering them, but by crossing off from each side considerations that seemed of roughly equal weight. What remained after the crossing-off, he wrote, showed him *"where the Balance lies,"* and if he had taken enough days and been through enough, that balance *"proved"* his decision.

Franklin's moral algebra is a remarkably sophisticated tool for its era. It incorporates several principles that modern decision science validates: the importance of generating options and evidence over time rather than in a single session (to allow incubation and to counteract availability bias); the explicit comparison of pro and con considerations, rather than focusing on one direction; the weighting of considerations by importance (even if informally); and the use of an external record that creates accountability and enables review.

Modern decision science improves upon Franklin's approach by addressing several key areas he couldn't have anticipated. While Franklin pioneered the use of external records and weighted comparisons, he was unaware of cognitive biases like anchoring, loss aversion, or confirmation bias, concepts that wouldn't be defined for another 200 years.

Today, we also incorporate base rates and reference class data to improve accuracy, alongside systematic tracking to create a continuous learning loop. Despite these advancements, the core of Franklin's 1772 method remains valid: using a structured format, considering clear alternatives, and allowing time for reflection are still the foundations of sound judgment.

Jim keeps a printed copy of Franklin's letter in his desk drawer. He shows it to new members of his leadership team when he explains the Decision Sciences culture at MegaHealth. *"This is not a new idea,"* he tells them. *"It is an ancient idea that most modern organizations have somehow managed to forget. We are remembering it."*

John Von Neumann, Game Theory, and the Art of Anticipating Others

No discussion of the intellectual foundations of Decision Sciences is complete without John Von Neumann, the Hungarian-American mathematician who, in 1944, co-authored with economist Oskar Morgenstern the Theory of Games and Economic Behavior, the founding text of game theory.

Von Neumann's achievement was to formalize the mathematics of strategic interaction, situations in which the optimal choice for one decision-maker depends on the choices of others, and to demonstrate that such interactions have determinate solutions under specific conditions.

The practical contribution of game theory to organizational Decision Sciences is not, primarily, the formal mathematical solutions. Most real organizational decisions are too complex and information-poor for the formal theory to apply directly. Its contribution is conceptual meaning that the habit of thinking about decisions as strategic interactions rather than as choices made in isolation.

Every significant organizational decision is made in a competitive, regulatory, or social environment in which other agents have interests and will respond to the organization's choices. The failure to model those responses, to think carefully about what competitors, regulators, partners, and patients will do in response to each option, is among the most common and most costly omissions in strategic decision-making.

Susan incorporates a simplified game-theoretic analysis into the Assess phase of ADEL for any decision with significant competitive implications. The analysis is not mathematically rigorous; it is a structured thought experiment that asks, for each strategic option under consideration: *"What is the most plausible response of the three or four most important other players? What does their response imply for the expected outcome of this option? And which option is most robust to the full range of plausible competitive responses?"*

This is the application of Von Neumann's fundamental insight, that rationality in competitive environments requires modeling others' rationality, to the practical conditions of healthcare strategy. It is not game theory in the full formal sense. But it is the cognitive habit that game theory prescribes, stripped of its mathematical apparatus and made accessible to practitioners who are excellent strategic thinkers but not mathematical economists.

The Confidence Calibration Exercise: Susan's Gift to Jim's Senior Team

Nearly six weeks after the Linac decision is logged in the Decision Record, Susan conducts an exercise with MegaHealth's nine-member senior leadership team. She calls it a confidence calibration session. It takes ninety minutes and produces, by unanimous agreement, the most productive discomfort any of them have experienced in a professional setting.

The exercise is simple in structure. Susan has prepared fifty factual questions, each with a verifiable answer, drawn from MegaHealth's operating environment: patient volumes, market share statistics, quality metrics, regulatory requirements, competitive intelligence. For each question, she asks participants not just to provide an answer but to provide a 90 percent confidence interval: a range within which they are 90 percent certain the true answer falls.

She then reveals the answers and counts how many of each participant's 90 percent confidence intervals contain the true answer. A perfectly calibrated respondent, one whose stated confidence accurately reflects their actual accuracy, should capture the true answer 90 percent of the time.

These findings align with general research on overconfidence. More importantly, they clarify the specific confidence patterns within the MegaHealth leadership team. Our analysis shows where team members are most likely to overestimate their accuracy and how these habits impact their decision-making.

Leader	Role	Questions Answered	90% CI Hit Rate	Calibration Assessment
Jim Johnson	CEO	50	62%	Significantly overconfident
Dr. Patricia Walsh	CMO	50	78%	Moderately overconfident
Marcus Chen	CFO	50	71%	Overconfident
Kelsey Henderson	Dir. Rad. Oncology	50	84%	Slightly overconfident
Maria Santos	NE Regional Dir.	50	58%	Significantly overconfident
Susan Myles	Consultant/ CDO	50	88%	Near-calibrated

The most important finding is not who is most overconfident but the pattern of overconfidence. Jim's overconfidence is concentrated in market and competitive assessments, areas where his experience is genuine, but where the environment has changed rapidly enough that his pattern library is partially outdated. Maria's overconfidence is concentrated in operational questions, areas where she has high experience, but where the organization's data systems have not historically provided her with reliable feedback on her assessments.

In ninety minutes, this calibration exercise creates a precise map showing where the leadership team's confidence outweighs its actual accuracy. This map highlights exactly where the decision-making process requires external reviews, more evidence, or a clear admission of uncertainty to avoid costly mistakes.

Kelsey Henderson's near-calibrated performance on her clinical domain questions confirms what Klein's research on naturalistic decision-making predicts: in domains where the feedback loop is fast, clear, and reliable, clinical outcomes in radiation oncology, expertise produces genuine calibration. In domains where the feedback loop is attenuated, strategy, competitive intelligence, market forecasting, even genuinely expert leaders are systematically overconfident.

CHAPTER FOUR:
A TALE OF TWO TEAMS

Why identical data produces radically different outcomes,
and what Sun Tzu, Tolstoy, and modern neuroscience tell us
about the difference

Tolstoy and the Fog of Organizational War

In War and Peace, Leo Tolstoy presents one of the most penetrating analyses of organizational decision-making ever written, even though the organizations in question are nineteenth-century armies, and the decisions being made involve the movement of troops rather than the allocation of capital.

Tolstoy's central argument about battle, made most explicitly in his portrait of the Russian general Kutuzov, is that the outcomes of large, complex engagements are determined far less by the strategic plans of commanders than by the accumulated decisions of thousands of individuals responding in real time to conditions that no commander could anticipate. Kutuzov's genius, as Tolstoy portrays it, is not strategic brilliance but the wisdom to understand this, to resist the temptation of elaborate plans that will inevitably be disrupted by contact with reality, and to trust instead in the resilience and adaptive capacity of the forces under his command.

The Napoleonic general, by contrast, in Tolstoy's telling, is addicted to the plan. He believes that reality can be controlled by sufficiently detailed strategy, and that the outcome of battle is primarily determined by the intelligence and will of the commander. This belief is not merely wrong; it is actively harmful, because it creates a command culture in which field commanders suppress information that contradicts the plan, avoid decisions that deviate from orders, and lose the adaptive capacity that real-time conditions demand.

The organizational parallels are direct and unsettling. We often see healthcare strategy sessions produce detailed plans that fail within three days of meeting operational reality.

In many cases, a decision fails not because of a poor strategy, but because the people executing it lack a framework to understand the reasoning behind their choices. When hundreds of individuals make small daily decisions without a clear strategic context, the original plan quickly falls apart.

Tolstoy's wisdom, and Kutuzov's, is the wisdom of Decision Sciences: design the decision architecture, not just the strategy. Trust the intelligence of the people closest to the work and build systems that allow their local knowledge to improve rather than contradict the organizational direction.

Sun Tzu and Competitive Intelligence

Sun Tzu's The Art of War, written approximately five hundred years before the Common Era in ancient China, contains what may be the oldest systematic treatment of organizational decision-making under competitive conditions. Its relevance to modern healthcare strategy is not merely metaphorical; its core principles map directly onto the competitive dynamics of healthcare market strategy with striking precision.

Sun Tzu's most famous principle, *"Know yourself and know your enemy, and you will not be defeated in a hundred battles"*, is, in modern decision science terms, a statement about the importance of calibrated self-assessment and accurate competitive intelligence. The first part, knowing yourself, is exactly what behavioral economics research shows organizations are worst at. They know their strengths with far less accuracy than they believe; they are systematically blind to their own cognitive biases; they systematically overestimate their organizational capacity.

The second part, knowing your enemy, is the foundation of competitive scenario planning: the deliberate, disciplined modeling of competitor behavior under various conditions, including the conditions created by your own strategic moves. Sun Tzu understood, two and a half thousand years before game theory was formalized, that competitive decisions are not made in isolation. They are made in anticipation of responses, and the quality of a competitive decision depends on the quality of the competitor model embedded in it.

> *"Strategy without tactics is the slowest route to victory. Tactics without strategy is the noise before defeat."*

-Sun Tzu, The Art of War

This principal maps directly onto the relationship between the strategic decision (the Decide phase) and the decision cascade (the Execute phase). The strategic decision without the cascade is a vision without implementation. The cascade without the strategic decision is implementation without direction. Both are required, and they must be connected, each cascade decision must be traceable back to the strategic rationale it is intended to serve.

The Experiment: Same Data, Two Teams

At the start of the third month into MegaHealth's Decision Sciences implementation, Susan designs an experiment. She presents the same analytical report, a ZIP-code-level heat map of emergency room overuse, produced by her team from three months of claims data to two separate regional leadership teams simultaneously, without telling either team that the other has received it.

The heat map is specific, quantitative, and actionable. It shows, with ZIP-code granularity, the locations within MegaHealth's service area where emergency room utilization is highest relative to population, adjusted for acuity. The top three ZIP codes show ER utilization rates between 180 and 210 percent of the adjusted regional average. Something is wrong in those communities, and the data is clear that it is wrong.

This map shows where resources are being used, but it doesn't explain why they are being used so much. To find the cause, you have to stop and investigate, a step Susan calls the 'diagnostic pause.

Team A: The Action Bias in Full Expression

Team A, led by the Northeast Regional Director, Maria Santos, is by any standard an excellent team. Maria is experienced, committed, and genuinely motivated by improving her community's health outcomes. Her team includes two operations analysts, a community health coordinator, and a finance partner. They are the kind of people who make things happen.

They receive the heat map on a Monday. By Thursday, Maria had convened the team and produced a recommendation. The analysis they have conducted is not trivial: they have looked at the affected ZIP codes' demographic profiles, reviewed transportation access data, benchmarked against similar communities that have deployed mobile health units, and developed a preliminary financial model and their recommendation is:

- Deploy a mobile health unit in the Riverside Heights ZIP code which is the highest-utilization area and provide primary care access for patients who currently lack transportation to fixed-site facilities.

  ★ Budget request:

    - $1.2 million.

  ★ Implementation timeline:

    - 90 days.

  ★ Expected impact:

    - 10-15 percent reduction in ER utilization within six months.

The recommendation is coherent, well-supported, and confidently presented. It is also built on an assumption that the team never explicitly examined: that the primary driver of ER overuse is inadequate primary care access attributable to transportation barriers.

This assumption was reasonable. Transportation barriers are a well-documented driver of ER overuse in underserved communities.

Maria's team had recent experience with exactly this pattern in a neighboring county, where a mobile health unit had dramatically reduced ER utilization. The availability heuristic, the tendency to judge probability by ease of mental access, was operating at full force. The transportation-access explanation was vivid, recent, and concrete. It felt like the answer.

The mobile health unit was approved. It launched on time. And three months after launch, ER utilization in Riverside Heights did NOT change.

The Root Cause That Nobody Asked For

The post-implementation analysis that Susan's team conducted in Month 4 produced findings that, in retrospect, were obvious but had been invisible to Team A because their decision-making process had not included a root cause analysis step.

Structured interviews with forty frequent ER users in Riverside Heights revealed a consistent pattern: 67 percent of the excess ER utilization was occurring between 6 PM and midnight, primarily for non-emergent pediatric illness. The patients were not coming to the ER because they lacked transportation to a primary care physician.

They were coming to the ER because their children got sick at seven in the evening, when every primary care office in the area was closed, and the mobile health unit, operating 9 AM to 5 PM, Monday through Friday, was as unavailable to them as any fixed-site clinic.

The solution was not mobile primary care. It was after-hours pediatric access. And it had nothing to do with transportation.

Team B: The Diagnostic Pause

Team B, led by the Southwest Regional Director, Marcus Webb, was a different kind of team. Not more talented than Team A. Not more experienced. But Marcus had been through a failure like what Team A would experience, a solution deployed without adequate root cause analysis that had consumed $800,000 and produced no measurable impact, and the experience had burned itself into his decision process.

When he received the heat map, Marcus's first question to his team was not *"what should we do?"* It was *"why?"*

They spent two weeks on the diagnostic question. They pulled the ER utilization data not just by ZIP code but by time of day, day of week, chief complaint, and patient age. They conducted ten structured interviews with frequent ER users. They reviewed the existing primary care access landscape, office hours, wait times, after-hours availability, for every primary care practice in the affected area.

The pattern that emerged was not what they expected. It was not transportation. There was no access to primary care during business hours. It was the specific, quantifiable absence of after-hours pediatric access. Families with children were making a rational decision: their child is sick at 8 PM, the pediatrician's office is closed until tomorrow, and the emergency room is the only option available. Given this, their recommendation was:

- Partner with three urgent care centers to extend pediatric evening hours to 10 PM, seven days a week.

    ★ Budget request:

        ▪ $340,000 - Less than a third of Team A's proposal.

    ★ Implementation timeline:

        ▪ 45 days.

    ★ Expected impact:

        ▪ 15-20 percent reduction in evening-hour ER utilization within three months.

The partnership was implemented. The ER visits in the target area dropped by 15 percent within three months. The freed capacity went to genuinely high-acuity patients who needed emergency care. The per-patient cost of the reduction was dramatically lower than any mobile health unit alternative.

The Action Bias: Why We Act Before We Think

Action bias, the systematic preference for action over inaction, even when inaction or a different action would produce superior outcomes, is one of the most consequential and least recognized cognitive biases in organizational settings. Research by Uriel Hahn and colleagues at Cardiff University demonstrated that people rate identical action and inaction differently, consistently attributing more agency, responsibility, and intelligence to decision-makers who act than to those who deliberate.

In sports, goalkeepers in penalty shootouts dive to the left or right approximately 94 percent of the time, even though remaining in the center provides the highest probability of stopping a goal. They dive because acting, even ineffectively, is psychologically preferable to appearing passive. In organizational life, the action bias produces decisions like Team A's: solutions deployed before problems are fully understood, because the appearance of decisive action is rewarded regardless of whether the action was the right one.

The Debrief: What Decision Sciences Reveals About the Gap

When Susan brings both teams together for a joint debrief four months after the experiment began, the atmosphere is initially charged. Maria Santos, Team A's director, has seen the outcome data. She knows her mobile health unit failed. She knows Team B's intervention succeeded at a fraction of the cost.

She is professional and gracious, but is covering something that looks a great deal like the particular combination of shame and defensiveness that arises when good people discover that they were not as thorough as they believed.

Susan addresses this directly at the start of the debrief and says:

> *"Before we analyze what happened, I want to say something that I mean sincerely. Team A did not fail because of incompetence. Team A's process would have succeeded in a situation where the transportation-access hypothesis was correct, and in many communities, it would have been. The problem was not the hypothesis. The problem was the absence of a step in the decision process that would have tested the hypothesis*

before committing resources to it. Team B did not succeed because Marcus is smarter than Maria. Team B succeeded because Marcus had a process that built in the diagnostic pause, the explicit question of 'why' that precedes the question of 'what'. That process advantage is something we can give to every team in this organization. That is what we are building."

Hearing this, Maria asks, *"Why didn't my team ask the 'why' question?"*

To this Susan responds, *"Three reasons. First, the availability heuristic: your team had direct experience with the transportation-access pattern in a neighboring county, and that experience made the transportation hypothesis feel correct without requiring examination. Second, the action bias: your team is results-oriented and solution-focused, which are generally virtues, but in this case those virtues created pressure to move from problem to solution faster than the evidence supported. Third, your decision process did not have an explicit diagnostic phase. You moved from 'here is the heat map' to 'what should we do' without a structured step that asked 'what is actually causing this.' No individual on your team was failing. The process was failing. And we can fix the process."*

The Feedback Loop: Where the Real Value Lives

The debrief reveals something else that Susan considers the most important finding of the entire experiment: Team A's mobile health unit, despite failing to reduce ER utilization, is not a total loss if the organization can learn from it.

The unit has produced four months of data. That data shows, with granularity that the original heat map did not have, the specific utilization patterns and patient needs in Riverside Heights. It has built relationships with community members that give MegaHealth credibility it lacked before. And it has demonstrated, through the failure of one intervention, what the actual problem is, a finding that has direct value for the redesign of the intervention.

Susan, *"In a learning organization, a failed decision is not a sunken cost. It is a tuition payment. You paid $1.2 million to learn something specific and important about your community's health needs that you could not have learned for $1.2 million any other way. The question is whether you extract the full value of that*

114

knowledge by incorporating it into your next decision, or whether you classify the project as a failure, quietly discontinue it, and retain none of the learning.”

Jim, *“Most organizations do the second thing?”*

Susan, *“Yes, most organizations do the second thing. That is why their decision quality does not improve over time. They make expensive mistakes and call them failures rather than investments in organizational knowledge. The Decision Record system we are building does not allow the second thing. Every decision, successful or not, produces a Learning Record that feeds into the organization’s decision intelligence database. Failed decisions, if we learn from them honestly, are often more valuable than successful ones.”*

The Quarterback Analogy: Building Organizational Playbook Intelligence

Jim processes the Team A/Team B story for a week before returning to Susan with a question that reframes everything they have been discussing.

Jim, *“I keep thinking about the Tom Brady analogy you used a few months ago. You said I need to be the quarterback, and my direct reports need to know what I’m thinking. But what I’m realizing is that Brady’s team doesn’t just know what he’s thinking. They know the playbook. They know the principles. So when the play breaks down and he improvises, the receivers are improvising in the same direction he is. That is what we don’t have. We don’t have a decision playbook.”*

Susan, *“That is exactly right. And it is one of the most important insights of organizational decision theory: the best organizations don’t just make good individual decisions. They build a decision culture so thorough that when individuals are operating autonomously, at three in the morning, in an emergency, in a situation that no protocol anticipated, they are still making decisions consistent with the organization’s strategic direction. Because they have internalized the principles, not just the rules.”*

Jim, *“How do you build that?”*

Susan, *“You build it the same way Brady’s team built it. Repetition, feedback, film review, honest assessment of what worked and what didn’t. You create a decision culture where tracking and reviewing decisions is normal, where people are not embarrassed to say 'my decision was wrong' because that honesty is valued and modeled at the top.*

You make Jim, as accountable for his decision quality as any analyst on the team. And you build the infrastructure that makes the feedback loop automatic rather than dependent on individual initiative."

Thirty Days Later: 100,000 Decisions

One month after MegaHealth activates its decision tracking system organization-wide, Jim receives a notification on his phone at 7:23 in the morning. The system has logged its 100,000th decision.

He is sitting in traffic on Oak Street, having, for some reason, turned left instead of right that morning, taking the longer route to the office, when the notification arrives. He stares at it for a long time.

100,000 decisions. In thirty days. Across an organization of 3,000 employees.

He calls Susan from the car.

Jim, *"How is this possible? 100,000 decisions in a month?"*

Susan, *"I have been waiting for this call. The number feels impossible because you are still thinking of 'decision' in the narrow sense, the big, deliberate strategic choice. But the system is capturing something broader: every time a manager makes a significant call about staffing, scheduling, patient placement, supply ordering, or protocol compliance, the system prompts them to log it. Most of these are small decisions. But they add up. And here is the most important thing: the more decisions people log and get feedback on, the more they want to make sure they are making the right ones. Decision quality is becoming a social norm, not just a management initiative."*

Jim, *"What do the first 100,000 tell us?"*

Susan, *"They tell us that we have about 40,000 decisions that are genuinely automated, routine, consistent, low variance. They are being made well, and the right response is to leave them alone. We have about 50,000 decisions in the middle range, some consistency, some variance, some patterns worth examining. And we have about 10,000 decisions that show sufficient variance and consequence to warrant systematic analysis. That is where the first wave of improvement will come from."*

Jim, *"What does the best decision-maker in the organization look like?"*

Susan *"That is not yet a meaningful question. We have thirty days of data. But I can tell you what I see in the early patterns. The best decision-making is not where you might expect it. It is not at the top of the org chart. It is in the clinical areas where feedback loops are fastest, where a decision is made and the result is known within hours or days. The slowest decision improvement is in strategy and capital allocation, where the feedback loop is eighteen months or longer. Which is exactly why those domains need the most deliberate Decision Sciences infrastructure, they have the least natural feedback to calibrate against."*

Mariana: The Patient in the Data

The best Decision Sciences frameworks are ultimately in service of something more important than organizational efficiency or competitive advantage. In healthcare, they are in service of patients, the people whose health outcomes are the downstream consequence of the thousands of decisions made daily in health system offices, conference rooms, and clinical areas.

Mariana Torres is forty-six years old. She works as a school custodian in Riverside Heights, raising two teenage children as a single parent. She has hypertension and pre-diabetes, conditions she manages imperfectly because the demands of her schedule make regular primary care appointments difficult. For the past three years, her primary point of contact with the healthcare system has been the emergency room, not for emergencies, but for the gaps: the refill she could not get from a doctor she could not see, the symptoms that were not quite alarming enough to interrupt her work schedule but concerning enough to bring her to the ER after her kids were in bed.

Mariana appeared in Susan's heat map. She was one of the data points in the Riverside Heights cluster. She was one of the reasons Team A deployed the mobile health unit, a unit that operated during the hours when Mariana was at work, and therefore never served her.

She was one of the reasons Team B extended the urgent care evening hours. And she used those hours. Twice in the first month: once for a routine blood pressure check and medication adjustment, once for her son's ear infection. Both visits resolved without an ER encounter that would have cost twelve times as much and produced ten times as much system friction.

Mariana is unaware that she was part of a data study or a scientific experiment. All she knows is that she can now get her blood pressure checked at 7:30 p.m. without missing work, which has clearly improved her life.

This is the downstream human consequence of the difference between Team A and Team B. Decision Sciences is not an abstract discipline. It is a mechanism through which better thinking produces better outcomes for specific, irreplaceable human beings. That is why it matters. That is why it is worth the investment of framework development, cultural change, and organizational courage that it requires.

James Reason and the Swiss Cheese Model of Organizational Failure

The British psychologist James Reason, in his foundational 1990 book Human Error, developed what has become the most influential model of organizational failure in the safety sciences: the Swiss Cheese model. Reason proposed that in complex systems, such as nuclear power plants, aviation, and healthcare, accidents do not result from a single catastrophic error. They result from the alignment of multiple smaller failures, each alone insufficient to cause harm, that happen to line up in a way that creates an unobstructed path to disaster.

In a complex system, every safety measure, like policies, checklists, and inspections, is like a slice of Swiss cheese. Each slice is imperfect and has holes. Usually, these holes don't line up, so if a mistake passes through one layer, it is caught by the next. However, when several small failures and human errors happen at the same time, the holes in every layer align. This allows a dangerous mistake to pass through every defense all at once, leading to a disaster.

The Decision Cascade that Susan builds for MegaHealth's CardioNorth initiative is, in Reason's framework, a system of multiple defense layers for strategic decision failure. Each layer, the Decision Record, the RACI matrix, the pre-mortem, the trigger-based review, is imperfect. Each has its own holes. But the combination of multiple layers, operating simultaneously and independently, dramatically reduces the probability that a fatal strategic failure trajectory can pass through all of them.

The organizational failure that Reason describes, multiple small cognitive and process errors aligning into a catastrophic outcome, is exactly what Team A experienced in the Riverside Heights mobile health unit decision. The availability heuristic (transportation model vivid and recent) aligned with the action bias (solution-focused culture under time pressure) aligned with the absence of a diagnostic pause (no formal step requiring root cause analysis before solution design) aligned with the authority of the Regional Director's recommendation (suppressing alternative perspectives). The holes lined up. The decision passed through all defenses unchallenged.

The Decision Sciences framework adds defense layers that break up the alignment. The pre-mortem would have asked: "We are deploying a mobile health unit and it has made no impact. Why?" The root cause analysis requirement would have forced the transportation assumption into the open before resources were committed. The trigger-based review would have flagged the flat utilization curve at Month 3 before Month 4's failure was confirmed. The holes might still have existed, but their positions would have been different, and the alignment would have been broken.

The Minority Report Problem: Where Decision Science Meets Ethics

As MegaHealth's decision system matures and the machine learning components become more sophisticated, a new class of decision begins to appear in Susan's weekly review queue.

These are decisions that the system is, in effect, recommending before the human decision-maker has initiated the decision process, cases where the predictive model has identified a pattern that historically precedes a specific decision, and is now prompting the relevant decision-maker to make the decision before the triggering condition has fully manifested.

The analogy to Minority Report, the Philip K. Dick story, adapted into the 2002 Spielberg film, about a police system that arrests people for murders they have not yet committed but that a predictive system has forecast, is not one that Susan introduces, but that Jim uses the moment Susan shows him the first example.

Jim, *"This is Minority Report. We are about to make a decision about a physician's performance before the performance problem has actually materialized."*

Susan, *"The analogy is apt. And it raises exactly the right question. The system has identified a pattern in Dr. Reyes's ordering behavior that, in 87 percent of historical cases, precedes a quality review that results in either coaching or remediation. The pattern is real. The predictive validity is genuine. The question is: what do we do with this prediction? Do we intervene before the pattern becomes a problem? Do we wait until the problem manifests? And if we intervene before, on what basis, and with what transparency to Dr. Reyes, do we justify the intervention?"*

Jim, *"The answer is not obvious."*

Susan, *"No. It is not. And I think the answer depends on what kind of intervention we are contemplating. If the intervention is a supportive conversation, 'We've noticed some patterns in your ordering that we'd like to understand better', that seems defensible and probably beneficial. If the intervention is punitive or restrictive, 'We are limiting your privileges based on a predictive pattern', that is ethically problematic. The decision system should enable the former and prohibit the latter. But we need to make this distinction explicit in the governance document, because the system itself doesn't know the difference. Only we do."*

This conversation initiates what becomes MegaHealth's Predictive Decision Ethics Protocol, a governance document that establishes clear principles for the use of predictive decision analytics in personnel, clinical, and operational contexts.

Its central principle: predictive analytics may inform and prompt decisions but may never constitute the sole basis for decisions that affect individuals' professional standing, patient care rights, or organizational accountability. The human judgment, informed by the prediction, is always the decision. The algorithm's role is input, not verdict.

*"The most dangerous thing we can do with decision intelligence
is treat algorithmic outputs as decisions rather than inputs to
decisions. The algorithm does not understand context. It does not
weigh competing values. It does not bear moral responsibility.
We do. The moment we forget that the moment we abdicate
our judgment to the system, we have built something that
optimizes for the wrong objectives with no one responsible for the
consequences."*

Susan, MegaHealth Predictive Ethics Protocol, Section 1

A Brief Interlude: Mariana's Year

What the data looks like from inside the data point

This is a story about one year in a patient's life. Their health was directly impacted by the decisions discussed in this book. It is told not just as a set of numbers, but as a human experience to show why these systems were created and to prove if they actually work.

January: The Emergency Room at 11:47 PM

Mariana Torres arrives at MegaHealth's Northeast Emergency Room at eleven forty-seven on a Tuesday night. She has been on her feet since five-thirty that morning. She worked an eight-hour shift at Lincoln Elementary, where she manages the custodial staff, then spent three hours cleaning two private homes for extra income, then made dinner for her sixteen-year-old daughter Camila and her fourteen-year-old son Tomas, then watched Tomas's left ear go from *"a little sore"* at dinner to *"I think he has a fever"* at ten-thirty.

She has no primary care physician who answers calls at eleven-thirty. She tried the urgent care clinic on Madison, which closes at nine. She tried the after-hours nurse line for her insurance, which suggested monitoring overnight and calling in the morning, advice that she found technically defensible and practically useless, because she has a nine-hour workday tomorrow and a fourteen-year-old who is crying into his pillow.

So she is here. The wait is forty minutes. The assessment finds acute otitis media, a garden-variety ear infection, completely treatable with a ten-day course of amoxicillin. The total cost of this encounter to the healthcare system: $847.

It is the third time Mariana has been in this emergency room in the past eighteen months. The first was for her own blood pressure, which was 168/104 at a workplace health screening and which her primary care physician, whom she sees perhaps twice a year, when she can schedule an appointment around her work hours, had been managing at a baseline that her recent readings suggest is no longer adequate. The second was for Camila's asthma inhaler, which had run out on a Sunday morning.

Mariana is not a frequent flyer, in the dismissive language that healthcare administrators sometimes use for patients who appear multiple times in the emergency room for non-emergent conditions. She is a working mother with inadequate access to the healthcare services she and her family need during the hours when they are available. She is not a system-abuser. She is a system-failure data point, someone whose needs are real, whose behavior is rational given her constraints, and whose presence in the emergency room represents a failure of access architecture rather than a failure of personal responsibility.

She is also one of the data points in the heat map that Susan shows Jim in the fourth month of their Decision Sciences work together.

March: The Mobile Health Unit Arrives

In March, a mobile health unit appeared in the parking lot of the Riverside Heights Community Center, two blocks from Mariana's apartment. It is there Tuesday through Thursday, nine to five. It offers primary care services, wellness screenings, and referrals.

Mariana passes it on a Wednesday morning, walking Tomas to the bus stop. She reads the sign. She notes the hours, nine to five, and calculates: she starts work at seven-thirty. She finishes at four-thirty, sometimes five-thirty. She has never, in the three weeks since the unit appeared, been able to use it. Some key points to consider here include:

- She does not know that the unit represents a $1.2 million investment by MegaHealth.

- She does not know that it was deployed based on the assumption that transportation barriers were the primary driver of her community's emergency room utilization.

- She does not know that she is, in some database somewhere, one of the data points that justified the deployment.

What she does know however is that it is not available to her, and that Tomas has had two colds and another ear infection since January, and that she has used the emergency room once more since her eleven forty-seven visit, for her own blood pressure, after a particularly stressful week in which she noticed that her vision was occasionally blurry.

June: The Evening Clinic Opens

In June, something different arrives. MegaHealth has partnered with the Riverside Urgent Care Clinic, which has been operating as a standard urgent care center, nine to seven, weekdays, to extend evening hours to ten PM, seven days a week.

Mariana learns about it from a flyer at Tomas's school, distributed as part of a school health communication campaign that MegaHealth's community health team has organized. The flyer says: *"Now open evenings and weekends for pediatric care."* It lists the address, the hours, and the services.

On a Thursday evening in late June, Tomas complains of ear pain again, his third episode in six months, which Mariana has been told may warrant referral to an ENT. She drives him to the Riverside clinic at seven-fifteen. They wait twelve minutes. A physician assistant examines Tomas, confirms a mild middle ear effusion rather than an acute infection, and provides a referral to a pediatric ENT whose office has Saturday morning hours.

The cost to the healthcare system for this was $184. A number, Mariana does not know, that is $663 less than the emergency room visit for the same clinical presentation in January.

The cost comparison misses an important detail from this visit. The physician assistant, Jessica, is fluent in Spanish and reviews Tomas's medical chart carefully. She asks detailed questions about how often his ear issues occur, whether they affect one ear or both, and his family history. She then orders a basic hearing test, which reveals a mild hearing loss in his right ear caused by the repeated buildup of fluid.

Nobody has ever tested Tomas's hearing before.

Mariana drives home at eight-forty-five, referral in hand, thinking about the parent-teacher conference in April where Tomas's English teacher had mentioned that he sometimes seemed distracted, or didn't respond when she spoke to him from the other side of the room. She had attributed it to the adolescent tendency to tune out teachers. She is now reconsidering.

August: Mariana's Own Health

In August, Mariana makes an appointment for herself. This is unusual. Her healthcare behavior, like that of many working parents in demanding jobs, has been organized almost exclusively around her children's needs. She gets her own care when circumstances force it, when a blood pressure reading is alarming, when a symptom is severe enough to interrupt her schedule. She has never made an appointment for preventive care.

She makes the appointment because the Riverside evening clinic is now a place she knows and trusts. Jessica recognized her when she came back with Tomas's follow-up for the ENT referral. The clinic feels, in a way that the emergency room never has and the twice-yearly primary care appointment never quite does, like a place that knows her.

The appointment reveals that her blood pressure is better controlled than it has been in three years, she has been more consistent with her medication since the visit where Jessica refilled it with a three-month supply and a clear explanation of why the consistency matters, but that her hemoglobin A1c is 7.2, borderline for the pre-diabetes that her primary care physician has been watching.

Jessica refers her to MegaHealth's diabetes prevention program, which offers evening sessions.

Mariana attends.

December: What a Year of Better Access Costs and Produces

At the end of the year, Mariana's family has had six encounters with the MegaHealth system that include:

- Two evening clinic visits for Tomas.

- One for Camila's asthma management.

- Three for Mariana herself.

Their total cost for this year is $1,247. What's worth noting here is that in the same period last year they had:

- Four emergency room visits.

- Two urgent care visits for the same family.

Their total cost last year was $4,836. That's considerably more than what they've paid this year, but the cost reduction is the least interesting thing that has changed.

Tomas has had tubes placed in his ears by the pediatric ENT. His hearing has returned to normal. His English teacher commented at the October parent conference that he seems to be paying much better attention in class, and that his reading comprehension scores have improved significantly. Mariana does not know whether the hearing loss was affecting his academic performance. She suspects it was.

Mariana's A1c is 6.7, down from 7.2, back in the normal range. She has not developed type 2 diabetes. The interventions that prevented its development were not dramatic: a twelve-session diabetes prevention program, a medication adjustment, and consistent access to the brief, quality primary care interactions that her schedule had previously made impossible.

She is a single patient, not just a number in a report. She is the inspiration for Susan's experiment, and the reason Jim writes a note on every decision file that says, *"This decision has a Mariana,"* reminding everyone that their choices affect real people.

The Connection Mariana Never Knows About

What Mariana Torres never knows, will probably never know, is that her health trajectory is part of a data set that researchers at MegaHealth's analytics team will eventually publish in the American Journal of Health Management. The paper describes the outcome differences between the mobile health unit deployment and the evening clinic partnership. It is careful, methodologically rigorous, and appropriately hedged about causal attribution. But its central finding is clear: the intervention that addressed the actual root cause of the access problem produced better health outcomes and lower costs than the intervention that addressed the assumed root cause.

The paper does not mention Mariana. It mentions *"patient cohort 7.B, working-age adults with pediatric dependents in the Riverside Heights ZIP code cluster."* Mariana is data point 23 in cohort 7.B.

The gap between data point 23 and Mariana Torres, between the health outcome measure and the actual human experience, is the gap that Decision Sciences, at its best, bridges. It builds the infrastructure that makes the right interventions more likely. It tracks the outcomes that indicate whether the interventions are working. But the outcomes are not numbers. They are nights when a parent drives her son to a clinic that is open, and a physician assistant speaks her language, and a hearing test reveals something that has been missed for months.

The goal of the decision database is to protect the unique and irreplaceable health of every individual. Susan, Jim, and the analysts all understand this. If the system is successful, they will continue to prioritize individual human lives, even as the amount of data increases and it becomes easier to focus only on broad statistics rather than the people behind them.

CHAPTER FIVE:
THE DECISION CASCADE

How one choice becomes forty-seven, and why the cascade determines whether strategy becomes reality

Nietzsche, Complexity, and the Will to Execute

Nietzsche's "will to power" is not about bullying or controlling others. Instead, it is the inner drive to master oneself, be creative, and constantly improve. In his book *Thus Spoke Zarathustra*, he introduces the "Overman," someone who imagines a better future and has the discipline to make it real. For Nietzsche, true power is the ability to shape your own life through dedicated action and self-control

Friedrich Nietzsche's concept of the will to power, often misunderstood as a crude endorsement of domination, is, at its deepest level, a description of the drive toward mastery, creativity, and self-overcoming that Nietzsche saw as the highest expression of human vitality.

In Thus Spoke Zarathustra, the figure of the Overman is someone who not only envisions a future but has the capacity to bring it into existence through persistent, creative, disciplined effort. The will to power, for Nietzsche, is not the desire to dominate others but the capacity to shape one's reality through committed action.

This is important to Decision Sciences because it identifies something most business books ignore: the gap between planning a goal and achieving it isn't usually caused by a lack of intelligence or data. Instead, it is a gap in execution. It is the willpower, discipline, and consistent systems that turn a smart decision into a successful result.

Most organizational strategies fail not because they were wrong but because they were not executed with sufficient discipline and adaptability. The Decision Cascade is the tool that bridges this gap, the mechanism by which a single strategic decision is unpacked into the full architecture of enabling decisions, assignments, timelines, and governance structures required to produce the intended outcome.

Nietzsche's will to power, operationalized: the decision cascade is how organizations exercise their will upon reality. Let's look at how this plays out at MegaHealth.

The CardioNorth Initiative: Building the Full Architecture

The CardioNorth Initiative is MegaHealth's decision to expand cardiology services into the Northeast Quadrant of its service area, an area with 340,000 residents, elevated cardiovascular risk, and inadequate cardiology access relative to population need. The strategic decision has been made: Option B, the Geographic Hedge, two clinics in geographically dispersed locations. What remains is the cascade.

Susan convenes the CardioNorth Decision Cascade session with a cross-functional team of twelve: representatives from real estate, physician services, clinical operations, finance, revenue cycle, information technology, marketing, and legal. Jim is present but is explicitly not the decision-maker in this session. Susan is the facilitator.

Susan, *"The purpose of this session is not to plan the CardioNorth implementation. It is to identify every decision that the CardioNorth strategy requires, assign a decision owner to each one, establish a decision timeline, and build the governance structure that ensures none of them falls through the cracks. We are going to map every decision, and I mean every significant commitment of resources, every significant choice among genuine alternatives, that flows from the strategic direction we have set. I estimate we will identify between forty and sixty decisions. Some of them will be visible immediately. Some will emerge as we work through the dependencies."*

Workstream 1: Real Estate and Facilities, Nine Decisions

The real estate workstream begins with what seems like a simple question: where, exactly, will the two clinics be located? This question immediately cascades into nine discrete decisions, each of which has genuine alternatives and genuine consequences.

The first decision, rent versus own, is a capital structure question with strategic implications beyond its financial components.

Owning creates a fixed asset on the balance sheet that constrains future flexibility but builds equity and eliminates lease risk. Renting preserves capital and flexibility but creates dependency on a landlord and exposes the organization to lease renegotiation risk at exactly the wrong moment (when the clinic has become established and the landlord knows it).

The financial analysis is necessary but not sufficient: the decision also depends on MegaHealth's long-term strategic commitment to the Northeast Quadrant, its current debt capacity, and its competitive vulnerability if a competitor occupies a key property before MegaHealth acts.

The fourth decision in this workstream, build-out standards for the patient experience, is one that most organizations treat as a facilities management question, but that Susan insists is a strategic brand decision. The patient environment in the new clinics will communicate something specific about MegaHealth's identity in the Northeast Quadrant.

The question isn't just about how the clinic looks, but also about the message it sends. Choosing that message requires a clear strategy that balances the organization's brand, the needs of the local community, and the strengths of nearby competitors.

The ninth and final decision in the real estate workstream, what happens to the leased space if the clinic is not financially viable by the end of Year 2, is one that no team ever wants to discuss. It is the sunset provision. Susan insists on it, because the organizations that do not plan for failure are the ones most likely to experience the sunk cost fallacy when failure arrives.

Workstream 2: Provider Strategy, Eleven Decisions

The provider strategy workstream contains the most consequential decisions in the CardioNorth cascade. Physician recruitment and retention is the primary success driver and the primary failure risk for the initiative, and it has the longest lead time of any workstream, requiring decisions to be made twelve to eighteen months in advance of clinic opening.

The first decision, employment versus independent contractor model for the lead cardiologist, is a classic *"it depends"* question that most organizations answer based on institutional habit rather than strategic analysis. Employment provides greater control over practice patterns, schedule, and quality protocols. Independent contractor arrangements offer potentially higher physician income and greater clinical autonomy, which can be recruitment advantages. The right answer depends on the specific candidate pool, the competitive landscape for cardiologist recruitment in the region, and MegaHealth's clinical management philosophy.

The third decision, compensation structure, involves a genuine strategic choice between base-heavy and productivity-heavy models, with different implications for physician behavior, patient volume, and quality outcomes. Research from the University of Washington shows that how doctors are paid changes how they treat patients, and these habits last for years.

To design a successful payment model, you need two different types of knowledge: an understanding of the job market and a clear medical philosophy. It is very rare for one person to be an expert in both areas.

The eleventh decision, what criteria will be used to evaluate the first year's clinical performance, and what will be done if those criteria are not met, is again the sunset provision. Susan has learned to include it in every high-stakes decision, because the willingness to pre-commit to clear evaluation criteria and clear consequences is one of the most reliable predictors of long-term decision quality.

Workstreams 3 through 8: The Full Architecture

The remaining six workstreams, Clinical Model (8 decisions), Financial and Revenue Cycle (6 decisions), Operations and Workflow (5 decisions),

Information Technology (4 decisions), Marketing and Growth (4 decisions), and Legal and Compliance (3 decisions), bring the total decision count to forty-seven.

The clinical model planning involves some of the most complex choices in the project. The first major decision is determining exactly which services to offer at opening. This is a difficult choice because each option significantly changes your needs for staff, office space, equipment, and your overall budget. Additionally, you must decide on the specific rules for how primary care doctors will transfer patients to CardioNorth cardiologists.

How will quality outcomes be measured and reported, and to whom? These are not operational details. They are strategic decisions that will determine whether CardioNorth is a low-complexity convenience clinic or a high-complexity specialty destination, and those are very different competitive positions requiring very different capabilities.

The information technology workstream contains what Susan considers the most underestimated risk in the entire cascade: the integration between CardioNorth's clinical systems and the MegaHealth enterprise EHR. Every EHR integration project she has observed in twenty years of healthcare IT has taken longer and cost more than projected. The base rate is unambiguous. The planning fallacy is fully operative. The corrective is to build the reference class data into the project timeline, assume 150 percent of the estimated integration timeline and plan accordingly.

Workstream	Decisions	Key Risk	Lead Time
Real Estate & Facilities	9	Lease terms / build-out delays	6-9 months
Provider Strategy	11	Physician recruitment timeline	12-18 months
Clinical Model	8	Scope creep / quality protocol gaps	6 months
Financial & Revenue Cycle	6	Payer contract timeline	4-6 months
Operations & Workflow	5	Staffing ramp-up coordination	3-4 months

Information Technology	4	EHR integration (typically 150% of estimate)	6-9 months
Marketing & Growth	4	Referring physician relationship development	6 months
Legal & Compliance	3	Certificate of need requirements	9-12 months

The RACI Matrix: Accountability Architecture

The forty-seven decisions of the CardioNorth cascade each require a governance structure that clearly assigns four roles that include:

- Responsible - who does the work to make this decision

- Accountable - who is ultimately answerable for the quality and timeliness of the decision

- Consulted - whose input is required before the decision is finalized

- Informed - who needs to know the outcome

The RACI matrix is not a project management formality. It is a cognitive intervention against the diffusion of responsibility, the well-documented tendency of individuals in groups to reduce their personal effort when responsibility is shared, on the implicit assumption that others are providing the effort. When no one is clearly accountable for a decision, the decision is at risk of not being made at all, or of being made by default, by the passage of time, by the accumulation of facts on the ground, by the path of least resistance rather than the path of greatest value.

The most common RACI error, in Susan's experience, is having multiple people listed as Accountable for a single decision. The RACI principle is absolute: every decision has exactly one Accountable. Multiple accountabilities are a euphemism for no accountability.

Susan, *"Jim, when I present the RACI for the first time, there will be pushback. People will argue that shared accountability reflects the collaborative nature of the organization. This is a rationalization. Shared accountability is a polite way of ensuring that someone else will handle it. I need your public support for the principle: one A per decision, no exceptions."*

Jim, *"Done."*

Decision Governance: The Review Cadence

Strategic decisions do not exist in a static world. Markets change. Competitors respond. Assumptions are violated. New information arrives. Effective decision governance includes a formal review cadence, a predetermined schedule and set of criteria for revisiting decisions in light of new evidence.

The CardioNorth governance structure includes five levels of reviews that include:

- Weekly operational reviews at the workstream level

- Monthly financial reviews with the CFO and CEO

- Quarterly strategic reviews with the executive team

- Annual comprehensive audits

- Trigger-based emergency reviews activated when specified threshold conditions are met.

The trigger-based review is the most innovative element.

For each of the forty-seven cascade decisions, the Decision Record specifies one or more triggers, specific, measurable conditions whose occurrence should prompt immediate reassessment. The Year 1 cardiology volume being less than 50 percent of projection at Month 6 is a trigger. A competitor announcing cardiology expansion into the target geography is a trigger. The lead cardiologist's resignation within the first twelve months is a trigger. These are not catastrophic events requiring panic; they are early warning signals requiring systematic reassessment of the decision architecture.

The trigger-based review system is the organizational equivalent of the circuit breaker in electrical engineering: a deliberate, pre-designed mechanism that interrupts a potentially destructive process before it reaches catastrophic failure. Most organizations allow strategic initiatives to fail over years because no one is formally authorized to say *"this is not working"* without the career risk of appearing defeatist. The trigger-based review removes that career risk by pre-authorizing the reassessment conversation and specifying exactly what conditions warrant it.

"The best decisions I have ever seen were not the ones made with the most confidence. They were the ones made with the most honesty about uncertainty, and with the most disciplined plan for detecting, as quickly as possible, whether the uncertainty was resolving favorably or unfavorably. The trigger-based review is how organizations stay honest over time." said Susan.

The Financial Architecture: Three Scenarios and the Lessons of Optimism Bias

No decision cascade is complete without a financial model, but Susan has learned to build financial models in ways that most healthcare finance teams do not.

Standard healthcare financial modeling is built around a single base-case projection, sometimes supplemented by a *"conservative"* scenario that is typically 20 percent below the base case. This approach has two problems that are:

- The base case is typically constructed from the inside view, from the optimistic details of the specific project rather than from the base rates of similar projects.

- A 20 percent haircut on an optimistic base case still produces an optimistic model.

Susan builds three scenarios for CardioNorth, and she builds them from the outside view first. The conservative scenario is explicitly designed around the base rate of cardiology expansion projects that have encountered headwinds such as:

- Weaker-than-expected payer mix

- Slower physician recruitment than projected

- A competitive response from the dominant regional health system

It is not 20 percent below the base case; it is what actually happens to cardiology expansion projects when two or three of the key assumptions prove incorrect.

Category	Conservative	Base Case	Optimistic
Year 1 Revenue	$1.8M	$2.4M	$3.1M
Year 3 Revenue	$4.2M	$5.8M	$7.9M
5-Year NPV	$3.1M	$5.7M	$9.3M
Break-Even	Month 38	Month 28	Month 22
Key Assumption	Recruitment 6-month delay; 40% payer mix	On-time recruitment; 55% payer mix	Recruitment advantage; 65% payer mix

The financial model reveals something that the base-case-only approach would have obscured: the CardioNorth initiative is NPV-positive even in the conservative scenario. This is a stronger justification for proceeding than the base case alone, because it means the decision is robust to at least one set of significant adverse developments. A project that is only viable in the base case is a project that requires everything to go right, which is the planning fallacy in its most dangerous form.

The Decision That Reveals the Leader: Risk Register and Sunset

The final element of the CardioNorth cascade is the risk register and sunset protocol. The risk register catalogs every significant risk that could impair the initiative, quantifies the probability and impact of each, identifies the decision owner responsible for monitoring and mitigating each risk, and specifies the trigger that would activate the contingency decision.

The sunset protocol specifies exactly what would constitute an insufficient performance to justify continuation of the initiative. This is the hardest conversation in every Decision Cascade session, and Susan never shortens it. The willingness to define failure conditions in advance, before any investment is made, before the sunk cost pressure is active, is one of the most reliable indicators of decision maturity in an organization.

Jim has been in this industry long enough to know that most organizations avoid this conversation. The standard rationale is that defining failure conditions is defeatist, or premature, or creates a self-fulfilling prophecy by giving people permission to give up. Susan's response to this rationale is consistently the same.

Susan, *"The organizations that avoid defining failure conditions are the ones most likely to experience them, because they have no early warning system, no authorized response, and no clear criteria for when the responsible thing is to redirect resources. The organizations that define failure conditions in advance are the ones that detect problems early enough to address them, course-correct efficiently, and preserve the capital and credibility to try again. Pre-committing to sunset criteria is not defeatism. It is the organizational equivalent of buying insurance before you need it, irrational to skip, and obviously right in retrospect when you do."*

The Commitment Problem: Ulysses, Self-Binding, and the Decision That Owns Your Future Self

In Book XII of Homer's Odyssey, Circe warns Odysseus about the Sirens, the creatures whose song is so beautiful, so irresistibly compelling, that sailors who hear it forget everything else and steer toward it, crashing their ships on the rocks. Circe's advice is practical: plug your sailors' ears with wax. But Odysseus, wanting to hear the Sirens' song without being destroyed by it, devises a different solution. He has his crew tie him to the mast and instruct them that, no matter how desperately he begs to be freed, they must not release him until the ship is safely past the danger.

This story, which the Nobel laureate economist Thomas Schelling made famous in his analysis of commitment devices, is one of the oldest illustrations in Western literature of a core Decision Sciences principle: the recognition that our future selves, under conditions of temptation, time pressure, or emotional distortion, will not make the same decisions our present selves would make under calm, deliberate reflection. The solution is not to trust the future self. It is to bind the future self, to create structures and commitments that constrain future decision-making in ways that preserve the values and intentions of the deliberate present self.

The sunset provision in MegaHealth's Decision Records is a commitment device. The pre-committed review trigger is a commitment device. The RACI accountability assignment is a commitment device. These are not administrative formalities; they are organizational equivalents of tying Ulysses to the mast, pre-commitments made when the sunk cost pressure is zero that constrain decision-making when the sunk cost pressure is at its maximum.

Behavioral economists Richard Thaler and Shlomo Benartzi used this idea to create the Save More Tomorrow program, the most successful retirement plan in U.S. history. Many employees did not want to lower their current take-home pay by putting more money into their retirement accounts. To solve this, they allowed workers to commit to saving in the future. Whenever employees received a raise, a portion of that extra money was automatically added to their retirement account before it ever reached their paycheck.

The commitment was made when the cost was hypothetical (a future raise not yet received) rather than real (current income). The result: a fourfold increase in retirement savings rates among participants. The behavioral intervention, not a change in incentives, not a mandatory requirement, just a commitment device that bound the future self using the present self's better intentions, produced an outcome that direct financial education and exhortation could not achieve.

The Decision Cascade's sunset protocol applies this principle to healthcare strategy. The commitment is made when the emotional cost of failure is hypothetical, before any resources are committed, before any relationship capital is invested, before any public announcements are made.

At that moment, the leadership team is in the same position as Ulysses before the voyage: able to hear the Sirens' song (the sunk cost argument for continuing a failing initiative) from a position of safety, and willing to bind themselves against it.

Clayton Christensen and the Innovator's Dilemma as a Decision Problem

Clayton Christensen's Innovator's Dilemma, the 1997 Harvard Business School Press book that Christensen himself described as the work he would most want to be remembered for, is, at its deepest level, an analysis of a specific and recurring pattern of organizational decision failure.

Christensen's observation was that successful, well-managed companies, companies that listened carefully to their customers, invested rationally in proven technologies, and made financially sound capital allocation decisions, were consistently disrupted by seemingly inferior technologies and products that their better-managed competitors had explicitly and rationally rejected.

The classic examples are well known. Disk drive manufacturers who saw hard disk size declining but rationally focused on their current customers' needs (capacity and reliability) rather than the emerging needs of new market segments (portability and price). Steel mills that produced high-quality steel for demanding customers who rationally rejected the lower-quality output of minimills, until the minimills improved their quality while retaining their cost advantage. Kodak, which invented the digital camera internally and rationally chose not to cannibalize its enormously profitable film business.

The Decision Sciences view of Christensen's Dilemma shows that the problem isn't poor strategy; it is the decision-making rules used to create that strategy. Established companies were using the right framework for their current market—listening to customers, checking financial returns, and investing in proven technologies. However, this same framework consistently led to the wrong decisions regarding how to survive long-term industry changes.

The correction is not to use a different strategy but to use a different decision framework for the different type of question: a framework that explicitly includes option value, the value of keeping strategic options open, and that accounts for the long-tail, low-probability scenarios where disruptive technologies become mainstream. The standard discounted cash flow model, which incumbents were using rationally and correctly, systematically undervalues options and catastrophically underweights low-probability/high-impact scenarios. Christensen's Dilemma is, in part, an artifact of the wrong decision tool applied to the wrong decision type.

For MegaHealth's CardioNorth decision cascade, Susan explicitly builds an options analysis layer into the financial model: rather than simply evaluating the three scenarios (conservative, base case, optimistic) for the full initiative, she values the option to abandon at each stage of the cascade, the *what is the value of being able to stop at Month 12 rather than committed to the full five-year program?"* question. This option value, consistently underestimated in standard capital allocation analysis, is part of what justifies the phased approach over a single large commitment.

The Site Visit: What Real Estate Reveals About Strategy

Three weeks after the CardioNorth cascade session, Jim drives out to the Northeast Quadrant on a Saturday morning. He does not tell anyone he is going. He takes no staff, no briefing materials, no data reports. He brings only the moleskin.

The drive from MegaHealth's administrative campus to the northern edge of the service area takes forty minutes on the highway and another fifteen on local roads. Jim takes the local roads. He passes through the commercial district that serves the corridor, a strip of nail salons, payday lenders, dollar stores, a single pharmacy, a family-owned Vietnamese restaurant that appears to be doing good business despite the early hour, and a CVS with a MinuteClinic. He notes the MinuteClinic. Competitor. Limited scope. But present.

He turns off the commercial strip into a residential neighborhood, the specific ZIP code cluster that the demographics analysis identified as the highest cardiovascular risk concentration in the service area. The houses are well-maintained, older. Many have vegetable gardens. Several front stoops display American flags and, in a few cases, flags of other countries, Guatemala, the Dominican Republic, Nigeria, a pattern of multi-generational immigration that the demographic data described but that looks different in person

He parks in front of a community center, a converted fire station, red brick, with a bulletin board by the entrance displaying notices in English, Spanish, and what he thinks might be Amharic. He sits in the car for a few minutes. He is doing something that no consultant's report does and that no dashboard represents: he is looking at the physical reality of the community he is proposing to serve.

The cardiology clinic he is planning to build, if the cascade decisions all go well, if the site is found, if the cardiologist is recruited, if the payer contracts are negotiated, will serve the people who live in these houses, walk this street, shop at that pharmacy, and apparently, on Saturday mornings, line up for something at the Vietnamese restaurant.

He writes in the moleskin: *"The restaurant is full at 9 AM. This is a community that uses shared spaces. The clinical model needs a waiting room that feels welcoming, not clinical. The community center bulletin board has three languages. Patient communication needs to go out in at least Spanish."*

He drives to two more candidate sites. At the second, a former bank branch with a parking lot, on a corner with good bus access and a crosswalk that connects to a residential street, he gets out of the car and stands on the sidewalk.

He thinks about Mariana. He does not know Mariana yet, she is still only data point 23 in cohort 7.B, but he is thinking about whoever she is. The working mother who cannot make a morning appointment. The diabetic patient who runs out of medication on a Sunday. The man with chest pain who calculates whether he can afford the emergency room.

He writes: *"This corner. The parking. The bus line. If I were a patient, not a CEO, a patient, working, with kids, and not enough money, and too much to do, this is where I would want the clinic to be."*

He photographs the corner. He photographs the bulletin board at the community center. He photographs the MinuteClinic entrance from across the street. When Susan and the real estate team present the final site recommendation four weeks later, these photographs are on the first slide of Jim's counter-presentation, before any financial analysis, before any demographic data. *"I visited the sites,"* he tells the team. *"I want to tell you what I saw, and then I want to hear what you recommend."*

This is unusual behavior for a CEO. It is also, Susan will tell him afterward, one of the most important things he has done in the CardioNorth process. The people who make the cascade decisions are more likely to make them well when the decision-maker has demonstrated that he understands the stakes at the human level, when the commitment is visible not just in the financial pro forma but in the forty minutes he spent standing on a street corner in a neighborhood he had never personally visited.

The Physician Who Said No

The provider strategy workstream of the CardioNorth cascade runs into its first significant obstacle in Month 4 of the cascade execution: the preferred candidate for the lead cardiologist position turns down MegaHealth's offer.

Dr. Ananya Krishnamurthy is forty-one years old, board-certified in interventional cardiology, and currently employed at the academic medical center in the city. She has been on the recruitment target list since Month 1 of the cascade, identified by Kelsey Henderson as the strongest available candidate in the regional market, with both the clinical credentials and the outreach skills that the CardioNorth model requires. She has had four conversations with MegaHealth's physician recruiter. She has toured the candidate site. She has had a two-hour dinner with Jim and the CMO.

Her letter, received by the recruiter on a Tuesday morning, is gracious and specific. She is declining, she writes, for three reasons. The first is the employment model: she prefers the independent contractor structure to the employment model that MegaHealth has offered, and while she understands the organization's preference for employment, the difference in compensation and autonomy is significant to her.

The second is the support infrastructure: she has specific concerns about the cardiology-specific IT support and the timeline for the new adaptive therapy equipment installation. The third is the geographic location of the candidate site: her family lives in the southwestern quadrant of the city, and the commute to the northeast site would add forty-five minutes to her daily drive.

Jim reads the letter twice. Then he calls Susan.

Jim, *"The Krishnamurthy recruitment has failed. I want to do a Decision Sciences post-mortem on the recruitment process."*

Susan, *"Good. What does her decline letter tell you about the assumptions in the Decision Record for that cascade decision?"*

Jim pulls up the Decision Record for the provider strategy workstream. He reads the key assumptions aloud.

Jim, "*Assumption 3: The employment model is acceptable to the target candidate pool. That assumption was wrong, or at least wrong for our top candidate. Assumption 5: The IT support infrastructure and equipment timeline are competitive with other regional employers. That assumption was also challenged. And the geographic assumption, we assumed the site location was acceptable to the candidate; we never asked directly.*"

Susan, "*Three assumptions violated in a single candidate's decline. Now the question: are these specific to Krishnamurthy, or are they systemic vulnerabilities in the CardioNorth recruitment value proposition?*"

Jim, "*I don't know.*"

Susan, "*That is the right answer. You don't know yet. But you now have a Decision Record gap that requires investigation before you recruit the next candidate. I want to propose that we run three structured conversations with cardiologists who are on the target list but have not yet been approached, specifically to test the employment model and IT infrastructure assumptions before they become recruitment obstacles.*"

The conversations happen. They confirm that the employment-versus-contractor preference is distributed roughly 60-40 in the candidate pool, a much wider distribution than the recruitment team had assumed. MegaHealth revises the cascade decision on employment model structure to offer both options, with different compensation frameworks, allowing candidates to self-select. The next candidate who reaches the offer stage, Dr. Marcus Reid, who is the third-ranked candidate on the original target list, accepts an independent contractor arrangement with a different compensation structure.

The Krishnamurthy rejection, logged and analyzed in the Decision Record system, has made the CardioNorth cascade more likely to succeed. This is the decision cascade working as designed: a failed enabling decision surfaces a violated assumption that, if uncorrected, would have become a pattern of failure rather than a single setback.

What the Financial Model Doesn't Capture: Community Trust

Five months into the CardioNorth cascade execution, Susan adds a metric to the program's success dashboard that was not in the original Decision Record. She calls it the community trust index.

The metric is built from a combination of sources: quarterly surveys of community members and primary care physicians in the Northeast Quadrant, social media sentiment analysis, and direct feedback from the community health workers who are the program's most direct link to the population it serves. It is imperfect, all its components are noisy and subject to interpretation, but it measures something that the financial model, the volume projections, and the quality metrics miss entirely: whether the CardioNorth clinic is becoming a trusted community institution or merely a new service in a zip code.

This distinction is vital for long-term success. Healthcare providers that act as trusted community hubs build strong relationships. Patients return, refer family members, participate in prevention programs, and rely less on emergency rooms because they trust these providers for their regular care. In contrast, services that are only convenient, offering good locations and hours but lacking genuine trust, only attract patients for occasional, one-time visits. This keeps them from building the long-term patient relationships necessary for a sustainable financial model.

Building community trust requires specific, intentional effort that does not appear in standard healthcare strategy planning. It requires clinical staff who look like and speak the languages of the communities they serve. It requires physical environments that feel welcoming to people whose experience of institutional spaces has not always been welcoming. It requires outreach strategies that go beyond marketing to genuine relationship-building with the community organizations, faith communities, and school systems that are the real social infrastructure of any neighborhood.

Jim assigns responsibility for the community trust index to the community health team, not to the marketing department, and includes it as a quarterly review metric at the same level of visibility as volume and revenue. The first quarterly report, at Month 6, shows a community trust index of 6.2 out of 10. The target for Month 18 is 7.5.

As Susan notes in her dashboard review, this is the only metric in the CardioNorth program that avoids the planning fallacy and optimism bias. You cannot calculate trust on a spreadsheet. Trust must be earned over time through consistent, honest interactions. You must be willing to acknowledge when you are failing to earn it, which allows you to change course when necessary.

A Check-In with The President Of The Board Elena

At Month 6 of the Decision Sciences implementation, Jim decides that Susan considers the most important signal of organizational transformation he has given: he presents the full decision analytics report, including the senior leadership team's calibration data, to the board's president Elena but in an official meeting. This is more like a check in.

Elena is a former healthcare executive turned venture capital investor, with a reputation for asking questions that no one else will ask and expecting answers that no one else can give. Elena has been a MegaHealth board member for three years. She was the one who pushed Jim most insistently to invest in the Omnimics data infrastructure, back when the rest of the board was skeptical of the cost. She has watched the subsequent Vairos implementation with approval.

And now, looking at the Decision Sciences learning report for the first time, she has a question that she asks with the careful precision of someone who has seen many organizational initiatives come and go.

This is not the edited version. Not the version that shows only the successes. The report shows the confidence calibration data, including Jim's own 62 percent hit rate on 90 percent confidence intervals. It shows the morning-versus-afternoon decision quality differential. It shows three decisions, identified by Decision Record number but not by decision-maker name, that had clear process failures, insufficient alternatives considered, key assumptions not surfaced, outcome metrics defined too loosely for meaningful evaluation.

There is complete silence in the room as Jim shares the report with Elena who takes her time to go over the report page-by-page. She's used to seeing boardroom material that presents the organization in the best possible light. But this time, the report is different and the uncertainty of how to respond to such data that is honest to the point of being uncomfortable is evidently clear by Elena's expressions.

Looking at Jim, she says, *"Jim, I have been on thirty-two boards and received thousands of board packages. This is the first time in my career that a CEO has shown me data indicating that he himself is overconfident about the things he knows least well. This is either the most reckless display of transparency I have ever seen, or the most sophisticated act of organizational leadership. I am trying to decide which."*

Jim, *"It is both. I am reckless enough to believe that this board can handle honest information. And I am sophisticated enough to know that organizations that do not show their boards honest information eventually produce crises that nobody saw coming, because everyone was managing the information to prevent discomfort rather than using it to prevent problems. We are building a decision culture here that depends on honest information flowing upward as well as downward. It would be inconsistent, and it would be wrong, to filter that information before it reaches the people who are responsible for governing this organization."*

Jim, Susan, and Elena produce three specific governance commitments that Elena drives through the discussion. These include:

1. A quarterly decision quality report to the board with the same transparency as the one Jim presented.

2. A formal board-level decision review process for decisions above $5 million capital commitment.

3. A board education session on cognitive bias and decision quality, to be facilitated by Susan, so that board members can apply the same critical lens to board-level decisions that the executive team is applying to operational ones.

It is, as Elena observes afterward in a private conversation with Jim, the first governance structure change at MegaHealth in eleven years that she genuinely believes will make the organization better rather than simply more compliant.

A Brief Interlude: Elena's Question What governance looks like when it is working

Before MegaHealth: The Board That Didn't Ask

Elena Vasquez's first board experience was in 2001, as an observer, she was then a senior vice president at a regional health system and had been invited to participate in a governance education program that included board observation. The board she observed was the governing body of a mid-sized community hospital that, within three years of that observation, would file for bankruptcy protection after a series of strategic miscalculations that included an ill-conceived ambulatory surgery expansion, an EMR implementation that consumed twice its budget and two years beyond its timeline, and a physician recruitment program that produced contractual liabilities the organization could not sustain.

What Elena remembered most about that board was not the specific decisions that proved wrong. It was the quality of the questions that were not asked in the boardroom. The ambulatory surgery expansion was presented with a single-scenario financial model. No one asked about the conservative case. The EMR was chosen following a vendor demonstration, and the board approved the budget without considering the organization's past performance on similar IT projects. Similarly, the physician recruitment program was approved based on presentations from enthusiastic staff, with no one requesting data on how many physicians from the last three groups stayed with the organization.

These were not exotic questions. They were the basic questions that any competent due diligence process would generate. They were not asked because the board culture did not reward asking them. The culture rewarded strategic optimism, forward momentum, and organizational alignment. Skeptical questions, questions that forced the examination of downside scenarios and past failures, were implicitly coded as disloyal or obstructionist. The board members who could ask them chose not to, calculating, correctly, that the social cost of asking exceeded the expected governance benefit.

Elena spent the following twenty years figuring out how to change that calculation.

What Elena Does Differently: The Question Protocol

Elena's governance philosophy has a specific and practical core: the systematic preparation and deliberate deployment of questions. Before every board meeting, she reads every document in the board package at least twice. In the first read, she marks anything she does not understand. In the second, she asks a different question: "What would I need to believe to find this convincing, and do I actually believe it?"

From these readings, she prepares her questions. Not all of them will be asked in the meeting, some will be answered by other discussion, some will turn out to be less important than they appeared on first reading. But she comes to every meeting with a written list, and she works through the list methodically, accepting answers that are genuinely responsive and following up on answers that are evasive, incomplete, or inconsistent.

This practice makes her, in the experience of every CEO she has served, simultaneously the most valuable and most demanding board member they have worked with. Most CEOs, once they understand what Elena's questions are doing, once they understand that she is testing the quality of organizational reasoning rather than expressing distrust of the leadership team, become advocates for her presence and her method. A few never quite make peace with the discomfort.

Jim belongs to the first group. He recognized, in the first board meeting where Elena challenged a financial projection with a specific question about the reference class for the projection, that she was doing something that his board needed: holding the organizational reasoning to the same standard of evidence and honesty that he was trying to build at the executive level. He has supported her presence and her questions ever since, partly out of strategic self-interest, her questions make his leadership team sharper, and partly out of something more fundamental: genuine admiration for someone who has built the governance practice that most boards describe and never quite achieve.

The Question That Changed the Governance Structure

Elena asks a question during this session that produces the most significant structural change in MegaHealth's board governance in a decade.

Elena Vasquez, *"Jim, I want to understand the composition of the board committees relative to the major decision categories that the Decision Sciences system has identified. Looking at your capital allocation data, the highest-risk decisions, the ones with the widest gap between projected and actual outcomes, are in technology and strategic partnerships. And when I look at our board committees, I notice that neither of those categories has a dedicated committee with the relevant technical expertise on it. We have a Finance Committee, a Quality Committee, and a Compliance Committee. We do not have a Technology Committee or a Strategy Committee. Do you think there is a governance gap there?"*

Jim looks at Susan, who has provided the decision analytics data that Elena is drawing on.

Susan gives the almost imperceptible nod that Jim has learned means: she is right.

Jim, *"Yes. There is a governance gap. The decision categories where our performance is weakest are not the ones where our board oversight is strongest."*

Elena Vasquez, *"Then I want to propose a governance restructuring. Two new committees: a Technology and Innovation Committee, with external members who have healthcare technology expertise; and a Strategy Committee, with external members who have regional healthcare market expertise. Both committees would review major decisions in their domains before they come to the full board, providing the pre-deliberation technical scrutiny that the current structure lacks."*

The proposal is adopted at the next board meeting. It represents something that Elena has been building toward for twenty years of board service: the application of Decision Sciences principles, the explicit alignment of decision oversight to decision risk, to the governance of an organization rather than just its operations.

Most governance reforms, Elena observes afterward, are reactive, they respond to failures, scandals, or regulatory pressure. This one is prospective, driven by data analysis of the organization's actual decision risk profile rather than by the most recent crisis. It is, she tells Susan, the most intellectually satisfying governance moment of her career.

CHAPTER SIX:
THE SCIENCE OF WISDOM

From individual cognition to institutional intelligence, and the
neuroscience of becoming genuinely better

William James and the Habits of Organizational Mind

William James, the nineteenth-century American philosopher and psychologist who effectively founded the discipline of psychology in the United States, wrote in his foundational 1890 text Principles of Psychology that habit is *"the enormous flywheel of society, its most precious conserving agent."* By this James meant that most human behavior operates through established habits, neurologically encoded patterns of response that are activated automatically by familiar stimuli, without conscious deliberation.

James's insight about individual habit applies with equal force to organizational habit. Every organization has a habitual response repertoire, standard ways of interpreting information, standard processes for making decisions, standard criteria for what counts as success. These organizational habits are often invisible to the organization's members precisely because they are habitual: they are the water the fish swim in, not something the fish can easily observe or analyze.

The most important organizational habits, in James's framework, are the ones that govern attention, what the organization notices, what it treats as signal versus noise, what it considers worth tracking. These attention habits determine, more than any other single factor, what an organization can learn from its experience. An organization that focuses only on financial metrics while ignoring patient experience will become highly skilled at optimizing finances but will remain blind to declining quality. Similarly, an organization that tracks outcomes without examining its decision-making processes will become an expert at justifying its performance yet will lack the ability to improve the thinking that drives those results.

Decision Sciences, in the language of James, is the practice of building better organizational habits of attention, habits that direct the organization's cognitive resources toward the decisions that most need scrutiny, the feedback loops that most need closing, and the learning opportunities that most need extraction.

Malcolm Gladwell and the Limits of Thin-Slicing

Malcolm Gladwell's 2005 book Blink popularized a concept he called thin slicing: the ability of the unconscious mind to find patterns in situations based on very narrow slices of experience. Gladwell's examples included a tennis coach who could predict a double fault before the server had finished the service motion, an art expert who knew immediately that a kouros sculpture was a forgery, and a couple's therapist who could predict the long-term outcome of a marriage from a fifteen-minute conversation.

Blink was widely misread as an endorsement of intuition over analysis, as a scientific argument that gut feeling is often more reliable than deliberate thought. This reading, which Gladwell has spent years trying to correct, misses the crucial qualification that makes thin-slicing valid in some contexts and dangerous in others.

Thin-slicing works when three conditions mentioned below are satisfied:

- The domain has stable regularities, meaning that the same cues reliably produce the same outcomes

- The expert has had extensive experience with many cases in that domain, providing a rich database of implicit patterns

- The feedback loops are short enough and clear enough that the expert has actually learned from experience rather than merely accumulated it

These conditions are present in some healthcare domains, the experienced emergency physician's rapid assessment, the master surgeon's feel for tissue tension, and absent in many others. Strategic decisions, capital allocation, market expansion, talent management: these domains are characterized by long feedback loops, noisy signals, and constantly changing environmental regularities.

In these domains, thin slicing, the executive's intuition that a market expansion will succeed, the board's instinct that a strategic partnership is right, is precisely the kind of judgment that behavioral economics research shows is most systematically unreliable. The lesson is not that intuition is worthless. It is that the value of intuition is domain-dependent, and the domains where organizational leaders most need help are precisely the ones where intuition is least reliable.

Six Months of Data: What MegaHealth Is Learning About Itself

Six months after the Decision Sciences system goes live at MegaHealth, Susan produces the first organizational learning report, a comprehensive analysis of what the decision database has revealed about MegaHealth's collective decision-making patterns.

The report contains findings that are simultaneously illuminating and uncomfortable. Jim receives it on a Friday afternoon and reads it twice before he calls Susan.

Jim, *"I need to understand the finding on page fourteen. The one about morning versus afternoon decision quality."*

Susan, *"We analyzed 6,200 significant decisions logged over six months, correlated against the time of day they were made, and then tracked the outcomes of those decisions over the following 90 days. Decisions made between 8 AM and noon showed outcome quality approximately 23 percent higher than decisions made between 2 PM and 5 PM, adjusted for decision type and decision-maker. That is a statistically significant finding across a large enough sample to be meaningful."*

Jim, *"Which confirms the decision fatigue hypothesis."*

Susan, *"Completely. But the more interesting finding is on page nineteen. Look at the pattern by decision type."*

Jim turns to page nineteen. The data shows that the morning-afternoon quality difference is largest for financial and strategic decisions, and smallest for clinical decisions. The clinical decisions show relatively stable quality across the day, with an exception at shift handoff times, where quality drops noticeably.

Jim, *"The clinical decision quality is stable because clinicians have external forcing mechanisms, protocols, checklists, structured handoff tools, that compensate for cognitive fatigue. The strategic and financial decisions don't have those forcing mechanisms, so they're fully exposed to the fatigue effect."*

Susan, *"Exactly right. And this tells us exactly where to invest next. We need to build the equivalent of clinical checklists for strategic and financial decisions, structured decision protocols that maintain quality even under conditions of cognitive load. The clinical world figured this out sixty years ago. The administrative world hasn't caught up."*

Atul Gawande and the Checklist Revolution

Surgeon and writer Atul Gawande's 2009 book The Checklist Manifesto documents one of the most powerful quality improvement interventions in modern medicine: the WHO Surgical Safety Checklist, a nineteen-item list that, when implemented in eight hospitals across eight countries, reduced surgical complications by 36 percent and surgical deaths by 47 percent.

The checklist works not because surgeons don't know the items on it, they know them all, but because cognitive load, complex environments, and the natural human tendency toward cognitive shortcuts produce reliable omissions that the checklist prevents.

The implication for organizational decision-making is direct: just as surgical teams benefit from structured checklists that enforce attention to critical decision points even under conditions of expertise and time pressure, strategic decision teams benefit from structured protocols that enforce attention to key questions, alternatives considered, assumptions surfaced, failure conditions defined, even under conditions of experience and urgency.

Elena Vasquez and the Board

At MegaHealth's quarterly board meeting in Month 8 of the Decision Sciences implementation, Jim presents the organizational learning report for the first time to the full board.

Elena Vasquez says "Jim, the report tells me what decisions were made and which ones produced good outcomes. What it doesn't tell me is whether the organization is actually getting better at making decisions, or whether you are just getting better at tracking and rationalizing the decisions you were already making.

The boardroom is quiet.

Jim, "That is a fair question. Susan, can you answer it?"

Susan is in the room for the first time at a full board meeting. She was not on the agenda.

Susan, "The honest answer is that we don't yet have enough longitudinal data to measure improvement in decision process quality as distinct from decision outcome quality. What we can measure is process adherence, whether decisions are going through the ADEL framework, whether alternatives are being documented, and whether assumptions are being surfaced. On that measure, adherence has improved from about 40 percent of significant decisions in Month 1 to 78 percent in Month 6. Whether that process improvement is producing better outcomes, we will be able to answer with confidence at Month 18. We do not yet know. And I think it is important to be honest about that rather than claiming results we don't yet have."

Elena continues "I am reading your report. On page seven, you say that the decision system has logged 100,000 decisions in thirty days. I want to understand: are you logging decisions that were previously being made implicitly, decisions that happened but were never recorded, or are you logging decisions that were previously not being made at all, and are now being made because you have created a process that requires them to be made?"

Susan, "Both, but in different proportions depending on the decision category. The clinical and operational decisions are primarily previously implicit decisions now being made explicit, they were happening before; we are now tracking them.

The strategic and governance decisions are a mixture: some were happening implicitly, some were genuinely being avoided, decisions were deferred or made by default because no one wanted to own them."

Elena Vasquez, *"The second category concerns me. If we are now making decisions that were previously not being made, if we are flushing avoidance behavior into the system, are we confident that those decisions should be made by the people making them? Or have we inadvertently created a system that produces decisional overload in the wrong places?"*

Susan is quiet for a moment. This is a question she had not anticipated, and it is a good one.

Susan, *"That is the most important governance question in the system, and I want to be honest with you: I don't have a complete answer yet. What I can tell you is that the system is designed with escalation logic, decisions above certain impact thresholds are automatically flagged for senior review, rather than being handled by whoever logged them. But your question suggests that the escalation logic may need to be more sophisticated, that we need to think not just about decision impact but about decision-maker authority alignment. I will have a proposal for you and Jim within two weeks."*

Elena Vasquez, *"That is the right answer. Not the reassurance, but the honesty and the plan. I appreciate it."*

The Neuroscience of Organizational Learning: What Changes When Decisions Improve

The research on individual learning provides a biological substrate for understanding organizational learning that is both rigorous and practically applicable. The brain's capacity to learn from experience, to revise its models of the world based on prediction error, is one of the most studied phenomena in neuroscience, and its principles translate surprisingly directly to organizational contexts.

Neuroscientist Wolfram Schultz, building on research by Peter Dayan and Read Montague, discovered that the brain's dopamine system functions as a prediction error signal. This system activates when results are better than expected, decreases when results are worse than expected, and remains steady when results match expectations. This gap between what the brain expects and what happens is how the brain updates its internal models.

Organizations learn through an analogous process, but with a critical vulnerability: the organizational prediction error signal can be blocked, suppressed, or misdirected by the same authority gradients, confirmation biases, and psychological safety deficits that distort individual decision-making. When the CEO's predictions are always confirmed (because no one will report the contrary), the organizational learning system is disabled. The leadership team that never receives accurate negative feedback is the organizational equivalent of a brain whose dopamine system has been disconnected from reality.

The Decision Sciences infrastructure that Susan builds at MegaHealth is, at a deep level, a mechanism for restoring the organizational prediction error signal. By creating a system in which decisions and their actual outcomes are systematically compared, and in which the gap between prediction and reality is surfaced, examined, and acted upon, the system restores the feedback loop that organizational learning requires.

Carol Dweck and the Growth Mindset Organization

Stanford psychologist Carol Dweck's research on what she calls the growth mindset versus the fixed mindset provides a useful framework for understanding the cultural dimension of Decision Sciences implementation. Dweck distinguished between people who believe their abilities are fixed traits (the fixed mindset) and people who believe their abilities can be developed through effort, good strategies, and input from others (the growth mindset). Her decades of research demonstrated that mindset predicts learning, resilience, and achievement in ways that measured intelligence and prior performance do not fully explain.

The growth mindset/fixed mindset distinction applies to organizations as well as individuals. Organizations with a fixed mindset about decision-making, those that believe their leaders are either good decision-makers or not, and that decision quality is a stable trait rather than a developing capability, are exactly the organizations that will not invest in Decision Sciences infrastructure. Why would you invest in improving a trait you believe is fixed?

Organizations with a growth mindset about decision-making, those that believe the quality of their collective judgment can be improved through deliberate practice, honest feedback, and systematic learning, are the organizations for which Decision Sciences delivers the greatest value. And they are, not coincidentally, the organizations that most rapidly develop the genuine organizational wisdom that compounds over time.

Jim to Elena Vasquez, after the board meeting, *"You asked whether we're actually getting better or just tracking better. I've been thinking about that question since the meeting. Here's my honest answer: I don't know yet. But here is what I do know. Three years ago, I would not have been able to ask the question clearly enough to answer it. I would not have had the framework to distinguish decision quality from outcome quality. I would not have had the data to even attempt the measurement. The fact that we can have this conversation, the fact that you can ask a question that sharp, and I can respond to it with something other than defensiveness, that is evidence of change. Whether it's the right kind of change, in the right amount, I will be able to tell you in twelve months."*

The Science of Organizational Learning: What Senge Got Right and What He Missed

Peter Senge's The Fifth Discipline, published in 1990, is one of my favorite books recommended by Senator Rick Scott while he was CEO at then Columbia/HCA. Senge's most widely read and most widely implemented framework for organizational learning ever produced. Senge's concept of the *"learning organization"*, a company in which people continuously expand their capacity to create the results they truly desire, where new and expansive patterns of thinking are nurtured, resonated so deeply with a generation of managers that it became the dominant framework for thinking about organizational change throughout the 1990s.

Senge's five disciplines, personal mastery, mental models, shared vision, team learning, and systems thinking, are genuine and valuable. In 1997, Harvard Business Review identified The Fifth Discipline as one of the seminal management books of the previous 75 years.

But the thirty years since The Fifth Discipline's publication have revealed a significant limitation in the framework: it is largely silent about the specific mechanisms by which organizations actually learn from decisions. It describes the culture and mindset of the learning organization, what it values, how it thinks, without providing the operational infrastructure for building it.

This is not a trivial omission. The research on organizational learning that has accumulated since 1990, from the work of Edmondson and Argris on defensive routines to the neuroscience of reward-based learning to the cognitive science of feedback loops, has made it clear that the intention to learn is necessary but not sufficient. What transforms intention into capability is operational infrastructure: specific mechanisms for capturing experience, comparing it to expectation, attributing the gap to specific causal factors, and feeding those causal conclusions into future decision processes.

Susan has designed MegaHealth's Decision Sciences system to serve as the operational infrastructure for the fifth discipline described by Senge. It does not replace Senge's cultural goals; instead, it provides the tools to put them into practice. True personal mastery becomes possible only when individuals receive accurate feedback regarding the quality of their decisions.

Mental models are made testable by explicit assumption documentation in Decision Records. Team learning is made systematic by the retrospective protocol. Systems thinking is made operational by the decision cascade mapping. And the learning loop, the mechanism by which each of these element's feed into improved future decisions, is built into the architecture rather than left to cultural aspiration.

The Jim-Susan Conversation That Changed Everything: Month Fourteen

At Month 14 of the implementation, Jim and Susan have a conversation that neither of them will forget.

It begins, as many of their most important conversations have begun, with something Jim noticed in the data. He has been reviewing the Decision Record database, something he now does for thirty minutes every Monday morning, and he has found a pattern that troubles him.

Jim, *"Look at this. The decisions that show the highest process compliance, the full ADEL protocol, every field completed, every step followed, are not the decisions with the best outcomes. In fact, some of our best outcomes over the past year came from decisions that went through an abbreviated process. And some of our worst outcomes came from decisions that followed the full protocol perfectly."*

Susan is quiet for a moment. She has seen this pattern herself. She has been thinking about it for several weeks.

Susan, *"You are about to discover the most important and most uncomfortable truth in all of organizational Decision Sciences. Process compliance is not the goal. Decision quality is the goal. Process is a means, not an end. A highly compliant process in the wrong context, a domain where the key variables are genuinely unknowable, or where speed genuinely is more important than analysis, or where the decision-maker has rare genuine expertise, can produce worse outcomes than the intuitive decision it was supposed to replace."*

Jim, *"So we built the wrong thing?"*

Susan, *"We built the right thing for the right purpose, which is to improve the quality of decisions in the domains where the framework adds value. What we have not yet done is calibrate the system to match the framework intensity to the decision type. We need to build what I call the Decision Architecture Matrix: a mapping of decision types to appropriate decision protocols. Some decisions need the full ADEL protocol. Some need an abbreviated version. Some should be delegated to structured decision rules. Some should be handled through recognition-primed decision-making by domain experts. The failure to differentiate is not a flaw in the framework. It is a calibration problem that we now have enough data to solve."*

The Decision Architecture Matrix becomes one of the most used tools in MegaHealth's Decision Sciences toolkit. It classifies decisions on two dimensions, decision reversibility and decision complexity, and maps each cell of the resulting matrix to an appropriate process. Reversible, simple decisions: standardized protocols or delegation. Irreversible, complex decisions: full ADEL with multi-day deliberation. Reversible, complex decisions: rapid experimentation and learning. Irreversible, simple decisions: structured decision rules with explicit override criteria.

This framework, which Susan draws from the broader decision literature and adapts for MegaHealth's context, is her answer to Jim's uncomfortable finding.

The data did not show that the ADEL framework was wrong. It showed that the ADEL framework was being uniformly applied to non-uniform decisions. The refinement makes the system more accurate, more efficient, and more respected by the clinical and operational leaders who had begun to chafe at applying full protocol to decisions that did not warrant it.

The Patient Whose Outcome Changed Everything: Maria's Redemption

Maria Santos, the Northeast Regional Director whose Team A deployed the failed mobile health unit, has spent twelve months rebuilding. She has been through the Decision Sciences training. She has applied the diagnostic pause to every significant decision in her region since the debrief. She has become, by any measure, one of the most rigorous decision-makers in MegaHealth's middle management tier.

In Month 15, her region identifies a new community health need: elevated Type 2 diabetes rates in the Southeast Quadrant, concentrated in a specific ZIP code cluster where primary care access is genuinely inadequate during business hours. The heat map is familiar. The problem is different.

Maria's process is completely different from the one that produced the mobile health unit. She spends three weeks on the diagnostic question before touching the solution space. She conducts fifteen structured interviews. She reviews three years of encounter data across twelve care settings. She talks with community health workers, pharmacists, and school nurses. These individuals interact with the population daily and possess a detailed, practical understanding of their lives and the obstacles they face when seeking care.

The finding that emerges is specific and actionable: the primary barrier is not access. It is medication adherence. Patients with diabetes in this ZIP code cluster have adequate access to primary care. What they do not have is adequate support for the complex medication management that Type 2 diabetes requires: understanding of the relationship between diet, activity, medication timing, and glucose control; a mechanism for monitoring their glucose levels without expensive equipment; and regular contact with a care navigator who can answer the questions that arise between appointments.

Maria's recommendation: fund a diabetes care navigation program, deployed through existing community pharmacies in the target area, at a cost of $180,000. Not a new facility. Not a mobile unit. A team of three care navigators, embedded in the community, connected to MegaHealth's primary care infrastructure by a shared data platform.

The program launches in Month 16. At Month 24, the outcome data is unambiguous: A1c control in the target population has improved by 0.7 percentage points, a clinically significant reduction associated with measurable decreases in diabetic complications. ER utilization in the target ZIP codes for diabetes-related complications is down 23 percent.

Maria mentions three specific patients she has met in person who avoided the dialysis that their Month-15 predictions indicated they would require. When Maria presents these results at the quarterly review, she pauses after showing the outcomes slides

Maria Santos, *"Three patients who are not on dialysis today. That is what this program means. Not the 23 percent utilization reduction or the 0.7 A1c improvement. Three people whose lives are materially different because we took four extra weeks to understand the problem before we deployed a solution."*

The room is quiet.

Jim, *"That is Decision Sciences."*

The Fourteen-Month Review: What the Data Actually Shows About Decision Quality

Susan's fourteen-month organizational learning report, the one she delivers to the board, the one that Elena calls the most honest board presentation she has received, contains seven major findings. Five of them confirm what the behavioral economics literature predicts. Two of them are genuine surprises.

- **Finding 1: Morning Decisions Are Materially Better** Decisions made between 8 AM and noon show 23 percent higher outcome quality than decisions made between 2 PM and 5 PM, adjusted for decision type, complexity, and decision-maker. This confirms the decision fatigue hypothesis and is consistent with the Baumeister-Danziger research tradition. Action taken: significant decisions rescheduled to morning blocks system-wide.

- **Finding 2: Written Pre-Deliberation Improves Group Decision Quality by 18 Percent** Decisions made in groups that used the pre-deliberation independent estimation protocol show 18 percent better outcome quality than comparable decisions made in groups without the protocol. The effect is strongest for financially complex decisions and weakest for clinical protocol decisions (where established clinical evidence provides a strong anchor that reduces the value of independent estimation). Action taken pre-deliberation estimation required for all Tier 3 and Tier 4 decisions.

- **Finding 3: The Pre-Mortem Has a Significant Asymmetric Effect** Decisions preceded by a pre-mortem show no improvement in outcomes when the decision is successful but show a 31 percent reduction in the severity of outcomes when the decision fails. The pre-mortem does not make successes more successful; it makes failures less catastrophic. This asymmetric effect is exactly what the theory predicts but is more dramatic than the research literature suggested. Action taken pre-mortem required for all decisions above $500,000 commitment.

- **Finding 4: Decision Record Completeness Predicts Outcome Quality** The completeness of the Decision Record, specifically, the number of alternatives documented and the specificity of the success metrics, is the single strongest predictor of outcome quality in the dataset, controlling for decision type, decision-maker experience, and resource commitment. Each additional alternative documented is associated with a 4.2 percent improvement in outcome quality. Each increase in success metric specificity (from vague to specific to quantitative) is associated with an 8.7 percent improvement. Action taken minimum standards for Decision Record completeness tightened.

- **Finding 5: Time-Pressured Decisions Show Systematic Optimism Bias** Decisions made under explicit time pressure, logged as "time-sensitive" in the Decision Record, show a 27 percent higher rate of unmet financial projections than decisions made without time pressure, controlling for decision type. The optimism bias is amplified by urgency. Action taken: explicit "time pressure debiasing" added to the Time-Sensitive decision protocol.

- **Finding 6 (Surprise): Women Decision-Makers Are Better Calibrated Than Men** Across the full dataset of 6,200 significant decisions, female decision-makers show 14 percent higher calibration accuracy (predictions vs. actual outcomes) than male decision-makers, controlling for decision type, experience level, and organizational tenure. This finding is consistent with a subset of the broader behavioral economics literature; several studies have found that women show lower overconfidence in domains outside their core expertise, but the magnitude of the effect is larger than expected.

Susan presents this finding with specific care, because it is sensitive and will be received differently by different people in the room. She is explicit about its limitations: the sample size for female senior decision-makers is smaller than for males, which may affect the reliability of the estimate; the finding is correlational and does not establish causation; and the individual variance within gender is far larger than the average difference between genders.

She is also explicit about its implication: the MegaHealth senior leadership team, which is 67 percent male, may be making decisions with a calibration disadvantage that a more gender-balanced team would not have. This is not an argument for tokenistic diversity. It is a genuine, data-driven observation about the relationship between team composition and decision quality.

- **Finding 7 (Surprise): Decision Quality Does Not Improve with Organizational Tenure** The most counterintuitive finding in the entire report: years of organizational tenure is not associated with improved decision quality. In fact, for decisions in fast-changing strategic domains (market competition, technology adoption), longer tenure is slightly negatively associated with outcome quality, consistent with the hypothesis that organizational insiders develop blind spots that make them less responsive to signals that contradict the existing strategic narrative.

The implication Susan draws carefully: this is not an argument for inexperienced leadership. Domain expertise is positively associated with decision quality in complex clinical and operational domains. But in strategic domains where the environment is changing rapidly, experience accumulated in a different environment may be a liability rather than an asset. The solution is not to replace experienced leaders with inexperienced ones. Instead, the approach should be to systematically expose experienced leaders to new perspectives that fall outside their current base of knowledge.

The Conversation Jim Has with His Leadership Team

Jim presents Susan's fourteen-month report to his full leadership team in a session he has specifically protected, three hours, no phones, no laptops except Susan's for the slides. He does not soften any of the findings.

The morning-versus-afternoon finding produces immediate practical discussion: three people volunteer to move their major decision meetings before noon without being asked. The gender calibration finding produces a longer silence and then a careful conversation about what it means and what it does not mean.

The finding about organizational tenure produces the most discomfort, partly because it directly challenges the assumption, universal among experienced executives, that their experience is an asset in all domains.

Dr. Patricia Walsh, CMO, *"The implication of the tenure finding, if I'm reading it correctly, is that I may be worse at strategic decisions in fast-moving domains than someone with less experience but more recent exposure to the relevant signals."*

Susan, *"That is one reading of the data, yes. Another reading is that the effect is domain-specific and that your clinical expertise, which is accumulated in a domain with reliable, fast feedback, is unaffected. It is specifically in the strategic domains where environmental change is fastest that the effect appears. And the correction is not to discount your experience but to systematically introduce decision inputs from people with different, and more current, reference frameworks."*

Dr. Walsh, *"Which means what, practically?"*

Susan, *"It means that your strategic decisions benefit from input from people who are newer to the organization, from external perspectives, and from data*

The conversation continues for ninety minutes. It is, by the agreement of every person in the room, the most honest and most productive discussion the MegaHealth leadership team has ever had about its own limitations.

CHAPTER SEVEN:
THE DECISION INTELLIGENCE INDUSTRY

Technology, consulting, academia, and the commercial ecosystem of better choices

The Market: From Philosophy to Platform

Decision Sciences has spent most of its history as an academic discipline, studied in university programs, published in specialized journals, practiced by specialized consultants serving primarily large corporations and government agencies. That is changing with extraordinary speed. The convergence of artificial intelligence, cloud computing, and enterprise data infrastructure is creating a commercial ecosystem for Decision Sciences that, within a decade, will be as ubiquitous in organizational life as customer relationship management software is today.

The global decision intelligence market, a term that encompasses the software, services, and analytics infrastructure that supports organizational decision-making, was valued at approximately $14 billion in 2023 and is projected to exceed $45 billion by 2030, growing at a compound annual rate of approximately 18 percent. Healthcare, financial services, and supply chain management are the three largest verticals by adoption rate, driven by the combination of high decision complexity, high decision stakes, and increasing data availability.

For healthcare leaders, this development is both a significant opportunity and a strategic necessity. Organizations that build decision intelligence infrastructure today will gain a compound advantage; each year of decision data makes subsequent choices more accurate. This advantage will be difficult, if not impossible, for competitors to replicate later. The window for a first-mover advantage in organizational decision intelligence is currently open, but it will not remain open indefinitely.

The Software Landscape

The commercial software landscape for decision intelligence divides roughly into three categories: decision support platforms, decision management systems, and decision intelligence platforms.

Decision support platforms provide analytical tools that inform decisions without managing the decision process itself. These include scenario modeling tools, simulation environments, optimization engines, and advanced analytics platforms. The vendors in this space include most major enterprise analytics providers, and the category is mature, competitive, and relatively well understood by enterprise buyers.

Decision management systems automate routine, rule-based choices that can be defined by clear "if-then" logic. These systems are commonly used for tasks like insurance claims, credit scoring, fraud detection, and managing medical orders. They are effective tools for handling simple or routine decisions, which allows human judgment to focus on more complex tasks where automation is not enough.

Decision intelligence platforms are the newest and most ambitious category: integrated environments that combine decision support, decision tracking, outcome measurement, and machine learning feedback loops in a single infrastructure. This is the category that Susan has been building from components at MegaHealth, and the category toward which commercial vendors are now racing. Leading platforms in this field are now offering integrated systems that connect decision-making to learning. Previously, this required expensive custom development. Now, these capabilities are available to organizations that cannot afford the high costs of proprietary development like MegaHealth.

Decision Sciences Consulting: The Service Ecosystem

The professional services dimension of the Decision Sciences market encompasses several distinct types of organizations. Global management consulting firms, McKinsey, Bain, Boston Consulting Group, all have developed Decision Sciences practices that focus on high-stakes strategic decisions, primarily for large corporate clients. Their methodologies are typically proprietary, their fees are substantial, and their value is concentrated in single high-stakes engagements rather than the long-term capability-building that Decision Sciences ultimately requires.

Specialized analytics and decision science firms, organizations like Decision Analyst, Mu Sigma, and dozens of regional boutiques, offer more focused services: quantitative market research, decision modeling, scenario analysis, and decision architecture design. These organizations are often better suited to the specific analytics work that feeds the Assess phase of the ADEL framework, and their services are increasingly accessible to mid-market organizations like regional health systems.

The most significant shift in consulting is the rise of behavioral science practices within traditional management firms. Major organizations like McKinsey's Center for Societal Benefit through Healthcare and Deloitte's Behavioral Insights practice, along with many smaller firms, now provide specific behavioral economics interventions. These include designing choice architecture, creating decision protocols, and offering training to reduce bias. These services go beyond simple data analysis to address the cognitive and cultural factors that determine decision quality.

The Academic Frontier: What Research Is Telling Us Next

The academic foundations of Decision Sciences are evolving rapidly in several directions, with direct practical implications for healthcare leaders.

The most significant development in decision neuroscience is the application of computational models, particularly reinforcement learning models from machine learning to the understanding of how the brain learns from experience. The work of Nathaniel Daw at Princeton, Samuel Gershman at Harvard, and colleagues has demonstrated that the brain uses algorithms remarkably similar to compthose of utational reinforcement learning to update its decision policies based on prediction err-r signals. This has two practical implications:

- It validates the design of decision tracking systems based on prediction error (comparing expected to actual outcomes)

- It suggests that the principles of effective machine learning, including the importance of exploration versus exploitation and the management of the credit assignment problem, may apply directly to the design of organizational learning systems.

In organizational behavior, the most significant recent development is the extension of psychological safety research from team to organizational and institutional levels. Amy Edmondson's work, and the work of her colleagues and students, has demonstrated that psychological safety is not merely a team-level phenomenon but can be cultivated at the organizational level through deliberate leadership behavior, structural interventions (such as anonymous reporting systems and protected dissent roles), and cultural norms that explicitly reward the surfacing of bad news.

In econometrics, the most practically important development is the increasing accessibility of causal inference methods to non-specialist practitioners. The work of economists Guido Imbens (Nobel Prize 2021) and Joshua Angrist (Nobel Prize 2021), along with Susan Athey at Stanford, has produced both methodological advances and accessible applications of causal inference that can be deployed by organizational analytics teams without requiring doctoral-level econometric expertise.

This development is critical for Decision Sciences because it enables organizations to move from correlation-based learning (we changed the protocol and outcomes improved) to causal learning (we have evidence that this protocol change caused the improvement), a distinction that makes organizational learning dramatically more reliable.

Netflix, Boeing, and the Cost of Not Deciding Well

Two contrasting case studies from outside healthcare illustrate the business stakes of decision quality at the organizational level.

Netflix's decision to greenlight House of Cards in 2011 has become the canonical example of data-informed bold decision-making. What is less often discussed is the decision architecture that produced it. Netflix's analytics team had identified a specific intersection of viewer preferences, high demand for political dramas, high demand for content featuring Kevin Spacey, high engagement with the British original, that supported a high-confidence prediction of audience interest in a US adaptation.

The decision to invest over $100 million was not primarily a creative judgment. It was the output of a decision process that had built a specific, falsifiable prediction and committed resources based on that prediction. When the show succeeded, the learning was not *"trust your gut on creative content."* The learning was *"the specific analytical model that predicted this audience intersection is valid. Build on it."*

The Boeing 737 MAX case provides the counterexample. The decision to modify the existing 737 platform rather than develop a new aircraft, driven by competitive pressure from Airbus and a desire to avoid the cost and timeline of type certification for a new aircraft, was a strategic decision that triggered a cascade of enabling decisions, each of which incrementally increased the safety risk of the MCAS software design. The failure was not a single bad decision but the accumulation of individually defensible decisions whose joint implications were never systematically assessed. The CASCADE, the forty-seven decisions in the language of MegaHealth's CardioNorth framework, was never mapped.

The individual decisions were each rationalized locally. Their systemic risk never surfaced organizationally. The result was two crashes, 346 deaths, and the largest aviation tragedy attributable to software in history.

Boeing's failure is a reminder that the Decision Cascade is not a bureaucratic formality. In safety-critical industries, and healthcare is among the most safety-critical, the failure to map, govern, and audit the full cascade of enabling decisions that flow from a strategic commitment is not merely an organizational efficiency problem. It is a patient safety problem.

"Every time I present the CardioNorth cascade to a team, I show them the Boeing case. Not to frighten them, but to show them the cost of treating enabling decisions as implementation details rather than as real decisions with real stakes.

The 737 MAX program had forty-seven decisions too. They just never mapped them.”, said Susan Myles.

The Conference: Decision Sciences Meets the World

In Month 16 of the MegaHealth Decision Sciences implementation, Susan is invited to present at the Behavioral Decision Research in Management conference, the leading academic-practitioner gathering for the field, held annually at rotating business schools. This year it is at the University of Chicago Booth School of Business.

She presents a forty-minute session on MegaHealth's implementation. She does not present it as a success story. She presents it as a twenty-month learning experiment, with honest reporting of what worked, what did not, what was harder than expected, and what the data actually shows.

The session produces more engagement than any other in the conference program. Not because MegaHealth's results are spectacular, they are encouraging but not transformative at the scale that healthcare leaders dream about, but because the presentation's honesty is unusual.

Most conference presentations of organizational Decision Sciences implementations describe the theory, the methodology, and the projected benefits. They do not describe the three months in Month 6 through 8 where adoption stalled, the senior leader who refused to use the Decision Record system for four months before finally complying, or the pre-mortem that revealed a $4 million assumption error in the CardioNorth model two weeks before the contracts were going to be signed.

After the session, Susan has conversations with practitioners from nine different organizations, a Fortune 500 pharmaceutical company, two regional health systems, a government agency, a technology firm, and four consulting practices. Every one of them is grappling with the same fundamental problem: how do you build the feedback loop between decision and outcome in a domain where the feedback loop is attenuated, noisy, and subject to confounding?

She also has a conversation with a researcher from the behavioral insights team at a major European government that will shape her thinking for the following year. The researcher, a young economist who has spent six years running randomized evaluations of government decision processes, sees that Susan carries back to MegaHealth and eventually incorporates into the second-year decision quality program.

Dr. Leila Mansouri, Oxford Behavioural Insights, *"The thing we keep finding in government contexts is that the biggest gains from Decision Sciences come not from improving the quality of individual decisions but from improving the quality of the decision to make a decision. Most organizations are making the wrong decisions because they are deciding the wrong things, because the decision agenda is determined by urgency and authority rather than by a systematic assessment of where better decisions would produce the most value. Your highest-impact decisions are probably not the ones that are getting the most analytical attention."*

Susan thinks about this on the flight back to Cincinnati. She thinks about MegaHealth's decision portfolio, the 6,200 decisions that have been logged and analyzed. She thinks about where the organization spends its decision-making energy and whether that allocation is optimal.

She thinks Dr. Mansouri is right.

When she gets back, she runs an analysis that she has not run before: not the quality of decisions that were made, but the quality of the decision about which decisions to make. Which decisions were given full ADEL treatment that were, in retrospect, routine enough not to need it? Which decisions were handled informally that were, in retrospect, consequential enough to deserve it?

The mismatch, the systematic misallocation of decision attention, turns out to be as important a source of organizational decision error as the biases and process failures that the existing framework was designed to address.

This insight becomes the foundation of MegaHealth's Year 3 decision quality program: not better frameworks for making decisions, but a better system for deciding which decisions deserve which framework.

Susan calls it the Decision Portfolio Review, a quarterly process in which the decision inventory is systematically examined for mismatches between decision importance and decision process intensity.

The Commercial Landscape in 2025 and Beyond

The Decision Sciences commercial market has accelerated significantly since Susan began her work at MegaHealth. Several developments are worth specific attention for healthcare leaders who are planning their own implementations.

The Integration of Behavioral Science into EHR Platforms

The major EHR vendors, Epic, Oracle Health (formerly Cerner), and Meditech, have all announced or deployed behavioral science features within their clinical decision support modules. These include choice architecture tools (default order sets designed based on behavioral economics principles), cognitive bias flags (alerts that surface when clinical ordering patterns suggest specific biases such as availability heuristic in diagnostic coding), and decision audit trails (structured logging of clinical decisions with their associated evidence, analogous to the Decision Record at the clinical level).

The integration of behavioral science into the EHR is significant because it places Decision Sciences tools at the point of care rather than in the strategy office. The challenge is that EHR-embedded behavioral science tools are subject to the same limitations as all EHR decision support: if they generate too many prompts, they create alert fatigue; if they are poorly targeted, they are ignored; and if they are designed without adequate clinical input, they impose cognitive burdens that reduce rather than improve clinical efficiency.

The Decision Intelligence Platform Category

A new category of enterprise software, Decision Intelligence Platforms, sometimes abbreviated DIPs, has emerged at the intersection of business intelligence, machine learning, and Decision Sciences theory. These platforms are designed to connect data, analytics, decision processes, and outcome tracking in an integrated architecture that approximates what Susan has built from components at MegaHealth.

The leading platforms in this category, which include offerings from established analytics vendors as well as purpose-built startups, vary significantly in their emphasis. Some are primarily analytical: strong forecasting engines, scenario modeling, and optimization capabilities, with limited decision governance features.

Others are primarily governance-oriented: strong decision record management, RACI functionality, and learning loop infrastructure, with limited analytical horsepower. The fully integrated platform, one that combines sophisticated analytics with robust decision governance in a single, clinically informed system, does not yet exist as a commercial product, which is why organizations like MegaHealth have been building from components.

Within two to three years, a fully integrated platform will almost certainly be commercially available. Organizations that have already established their foundational infrastructure, including a decision-making culture, habits for keeping decision records, and disciplined learning loops, will be ready to use this platform immediately. Conversely, organizations that have not yet done this will face a much harder double challenge: they will have to implement the new technology while simultaneously trying to build the necessary organizational culture to support it.

Academic Programs: The Next Generation of Decision Scientists

The academic pipeline for Decision Sciences practitioners is expanding. The University of Pennsylvania's Master of Behavioral and Decision Sciences program, founded in 2016, is now producing graduates who are being placed in behavioral science roles in healthcare, financial services, government, and technology. Harvard's Decision Science Laboratory, MIT's Laboratory for Decision Science, and Carnegie Mellon's Center for Behavioral Decision Research are producing doctoral graduates who increasingly move into industry roles rather than traditional academic careers.

The most significant development in healthcare Decision Sciences education is the emergence of Decision Sciences modules within healthcare MBA and MHA programs. Institutions including Duke, Johns Hopkins, Minnesota, Georgetown, and Belmont University have incorporated behavioral decision theory, decision process design, and organizational learning methodology into their graduate healthcare management curricula, reflecting a recognition, by healthcare management educators, that the discipline addresses a genuine and growing need in the leadership development of healthcare executives.

For MegaHealth's leadership development program, which Susan is now helping to redesign for Year 3, this academic pipeline creates an opportunity: recruiting graduates from behavioral and decision science programs into organizational roles that did not previously exist, and building the internal Decision Sciences expertise that currently relies heavily on Susan's individual knowledge.

The International Dimension: What Other Countries Are Getting Right

The United States is not the leader in applying behavioral science and Decision Sciences principles to healthcare policy and administration. That distinction belongs, in most comparative assessments, to the United Kingdom, which in 2010 established the world's first government-level behavioral insights team, the so-called "Nudge Unit", within the Cabinet Office.

The UK behavioral Insights Team, now a social purpose company that works globally, has applied behavioral science to healthcare challenges including organ donation (the switch to a presumed consent system that increased donation rates significantly), medication adherence (text message interventions that improved adherence by 25 percent in clinical trials), and screening participation (choice architecture changes to cancer screening invitation letters that increased response rates by meaningful margins).

Denmark, the Netherlands, and Australia have established similar institutional mechanisms for applying behavioral science to public health and healthcare policy. The common thread across these international examples is institutionalization: behavioral science is not deployed as a one-off consulting engagement but as a standing capability within the health system's governance structure, continuously applied to the full range of policy and operational decisions.

This is precisely the model that Susan is building at MegaHealth: not a periodic behavioral science assessment, but a standing Decision Sciences capability that is continuously engaged across the organization's decision portfolio. The international evidence suggests that this institutional model, behavioral science as an embedded function rather than an episodic intervention, produces benefits that compound over time in ways that periodic engagements cannot.

CHAPTER EIGHT:
THE AI INFLECTION POINT

Artificial intelligence, decision automation, and the governance imperative for healthcare leaders

Alan Turing's Question, Reconsidered

In 1950, Alan Turing posed the question that would define the next seventy years of computing: *"Can machines think?"* His paper in Mind, titled *"Computing Machinery and Intelligence,"* proposed what he called the imitation game, the test that has come to be known as the Turing Test, as a practical approach to the question. If a machine could conduct a textual conversation indistinguishable from a human's, Turing argued, the question of whether it *"really"* thought was no longer interesting. Its behavior would be equivalent to thinking for all practical purposes.

Turing was asking, fundamentally, a Decision Sciences question. The capacity to make decisions, to interpret information, generate options, evaluate alternatives, and commit to courses of action, is the operational definition of intelligence as Turing understood it. And the question of whether machines can participate in that capacity, and under what conditions and with what governance, is now the most urgent practical question in organizational Decision Sciences.

The year that Jim and Susan begin their Decision Sciences program at MegaHealth is also, not coincidentally, a period of dramatic acceleration in artificial intelligence capabilities. Large language models are being deployed in clinical settings for documentation, diagnosis support, and patient communication. Predictive analytics systems are moving from advisory roles to automated decision execution. The administrative and clinical decision landscape of healthcare is changing faster than the governance frameworks that are supposed to guide it.

This chapter is about that change, and about the specific governance and ethical challenges it creates for healthcare leaders who are serious about Decision Sciences.

Four Waves of Decision Automation in Healthcare

The automation of decisions in healthcare has proceeded in four distinct waves, each more powerful and more consequential than the last.

- **Wave One: Protocol Automation (1980s–2000s)** - Clinical pathways, order sets, and standardized treatment protocols codified clinical best practices into structured rules that reduced the cognitive load of routine decisions and reduced variation. The benefits were significant: evidence-based protocols, consistently applied, improved outcomes for common conditions. The limitations were equally significant: protocols are good at handling what they anticipate and brittle in the face of the unexpected.

- **Wave Two: Decision Support (2000s–2015)** - Electronic health records with embedded clinical decision support tools provided real-time alerts, reminders, and recommendations at the point of care. The evidence on the impact of clinical decision support is substantial. well-designed alerts reduce medication errors, improve protocol compliance, and surface diagnostic considerations that clinicians miss. The limitation: alert fatigue. When decision support systems generate too many alerts, particularly when alert accuracy is insufficient, clinicians habituate to them, clicking through the warnings without processing them. The decision support becomes invisible.

- **Wave Three: Predictive Analytics (2015–2022)** - Machine learning models trained on large clinical datasets began producing predictions about individual patient risk, sepsis early warning systems, readmission risk scores, deterioration alerts, that outperformed clinical judgment in controlled studies. The transition from protocol (if-then rules) to prediction (probability estimates) was intellectually significant: predictions can be calibrated, updated, and improved in ways that static protocols cannot. But predictions require interpretation, and the translation from "the model gives this patient a 73 percent sepsis risk score" to "what the clinical team should do" remained a human judgment.

- **Wave Four: Generative AI and Autonomous Agents (2022–present)** - Large language models and AI agents capable of synthesizing complex information, generating clinical notes, answering patient questions, and in some deployments initiating clinical workflows are transforming the relationship between human judgment and automated action in ways that the previous three waves did not. The critical distinction of Wave Four is the collapse of the gap between advice and action: where previous waves produced recommendations that humans then acted upon, Wave Four systems can increasingly act, draft the order, send the message, initiate the referral, leaving the human in a supervisory rather than a primary decision role.

The Automation Bias Problem

Automation bias, the tendency of human operators to over-rely on automated systems, accepting their outputs without adequate scrutiny, is one of the most well-documented and most consequential failure modes in human-automation interaction. This issue has caused aviation accidents, nuclear power plant incidents, and countless medical errors. These medical events often go unreported because they happen in situations where both the human and the automation are involved, making it genuinely difficult to determine exactly who or what caused the mistake.

The paradox of automation bias is that it is most severe precisely when automation accuracy is highest. When an automated system is correct 99 percent of the time, the human operator has very little incentive to maintain the vigilance required to catch the 1 percent of cases where it is wrong, and the skills and mental models required to catch those errors atrophy through disuse.

The result is a system that is excellent in routine conditions and vulnerable at exactly the moments when human oversight is most critical.

For healthcare leaders considering Wave Four AI deployments, automation bias is the governance challenge that most urgently requires attention. The issue is not whether AI can provide accurate clinical recommendations, in many areas, it is clear that they can. The challenge is whether the humans responsible for overseeing these recommendations are maintaining the level of active, skeptical engagement necessary to identify the errors that AI systems will inevitably commit.

Susan's governance framework for AI-augmented decision-making at MegaHealth uses four structural mechanisms to prevent automation bias. These include:

- Mandatory explanation requirements - AI-generated recommendations must always include an explicit statement of the evidence and reasoning on which they are based, requiring the human recipient to engage with the reasoning rather than simply accepting the conclusion.

- Calibration transparency - AI systems must display their historical accuracy for the specific type of recommendation in the specific clinical context, enabling the human to apply appropriate skepticism based on the system's actual reliability profile.

- Regular dissonance exercises - clinical teams periodically review cases where their AI-supported decision differed from what the AI recommended, maintaining the cognitive skills required for independent judgment.

- Explicit override culture - overriding an AI recommendation is treated as a legitimate and important clinical act that requires documentation and is reviewed for learning, rather than being treated as a deviation to be minimized.

The COMPAS Recidivism Algorithm: What Healthcare Can Learn from Criminal Justice

The COMPAS algorithm, a commercial risk assessment tool used by courts in the United States to estimate the probability of criminal reoffending, became the subject of a landmark investigative report by ProPublica in 2016.

The report found that the COMPAS algorithm exhibited a statistically significant racial bias. Specifically, it systematically overestimated the recidivism risk for Black defendants while underestimating it for white defendants. This resulted in disparate error rates across groups: Black defendants were more likely to be incorrectly flagged as high risk (higher false positive rate), while white defendants were more likely to be incorrectly labeled as low risk (higher false negative rate). These discrepancies had direct, measurable consequences for sentencing and parole outcomes. The COMPAS case illustrates three principles that are directly applicable to healthcare AI governance. These include

1. Algorithmic systems trained on biased historical data reproduce and amplify that bias.

2. The opacity of proprietary algorithms makes bias difficult to detect and impossible to fully audit.

3. The deployment of automated decision systems in high-stakes, legally consequential domains without adequate governance frameworks produces harms that are both significant and extremely difficult to remediate after the fact.

Healthcare AI systems trained on historically biased clinical data, where, for example, Black patients have been systematically undertreated for pain, face exactly analogous risks. The governance framework must address training data bias before deployment, not after harm has occurred.

Susan's AI Governance Taxonomy

Susan presents MegaHealth's AI governance framework to Elena in a session that she has specifically requested. The framework classifies AI applications in healthcare into four risk tiers and specifies governance requirements for each.

Risk Tier	Definition	Example	Governance Requirement
Tier 1: Informational	AI provides information that humans use to make decisions; no automation of decision execution	Diagnostic image pre-screening; clinical literature search	Standard accuracy validation; transparency disclosure to users
Tier 2: Advisory	AI provides specific recommendations with explicit reasoning; human makes final decision	Sepsis early warning; readmission risk scoring	Prospective accuracy validation; override documentation; bias auditing
Tier 3: Supervisory	AI executes decisions within defined parameters; human monitors and can override	Automated medication dose adjustments; scheduling optimization	Real-time monitoring; override culture protocols; regular human-AI accuracy comparison
Tier 4: Autonomous	AI executes decisions independently; human reviews post-hoc	Autonomous prior authorization; unmonitored patient communication	Prohibited without explicit regulatory approval and ethics board review

Elena Vasquez, *"I want to understand the rationale for prohibiting Tier 4. There will be pressure, financial pressure, to automate decisions wherever automation is possible. What is the principled case for drawing the line at autonomous execution?"*

Susan, *"The principled case is the same one that applies to any situation where the consequences of error are irreversible, and the affected party is a human being with rights and interests that the automated system is not constitutionally capable of understanding or weighing. Prior authorization decisions affect patient access to care. Fully autonomous execution of those decisions, without human review of individual cases, removes a check that exists for reasons beyond efficiency. It exists because the decision involves a human being whose specific circumstances, values, and medical history may not be fully captured in the data that trained the model. The automation's speed is not a sufficient justification for removing that check. The speed merely makes the potential for harm systematic rather than individual."*

Elena Vasquez, "And if a competitor deploys Tier 4 and achieves a 20 percent cost reduction that makes our cost structure noncompetitive?"

Susan, "Then we face a genuine strategic dilemma, and we should discuss it as such, transparently, with full information about the risks and the competitive stakes, rather than allowing the competitive pressure to quietly erode governance standards that we established for good reasons. The question you are asking is important and should be answered, not managed away. My answer, for the record: a health system that reduces costs by 20 percent through governance compromises that harm patients is not competing. It is extracting value from the most vulnerable party in the system. That is not a competitive strategy I would recommend, regardless of the financial pressure."

This session produces a unanimous resolution to maintain the Tier 4 prohibition, with an annual review clause that requires the board to explicitly reaffirm the policy considering evolving competitive and regulatory circumstances. The resolution is formally documented in MegaHealth's Decision Record as a governance decision. This record includes the pre-mortem analysis, a summary of the alternatives considered, and a defined sunset trigger, which outlines the specific regulatory changes that would require a formal reassessment of the decision.

The Promise and the Peril: What AI Does Well in Decision Sciences

The governance framework is not a prohibition on AI in Decision Sciences. It is an attempt to deploy AI where it genuinely improves decision quality and limit it where it creates governance risks that outweigh the efficiency benefits. In many areas of Decision Sciences, AI is genuinely transformative. Some of these areas include:

- **Pattern recognition at scale for example:** The most powerful application of AI in organizational Decision Sciences is the identification of patterns across large numbers of decisions that would be invisible to human analysis. The decision database at MegaHealth contains, after two years, more than three million decision records. No human analyst can examine three million records and identify the subtle patterns, the specific combinations of decision context, decision process, and decision-maker characteristics that predict outcome quality. Machine learning models can identify these patterns. They can pinpoint which decisions are consistently too optimistic, which types of decisions result in the most variation between different people, and which parts of a Decision Record best predict high-quality outcomes. This information provides actionable intelligence that the human Decision Sciences team can then use to improve organizational processes.

- **Reference class augmentation:** Building reference classes manually, identifying the set of past decisions that are most similar to the current one for forecasting purposes, is time-consuming and subject to selection bias. AI systems trained on the decision database can identify reference classes automatically, surfacing historical analogs that human analysts might not have found and quantifying the statistical similarity of current decisions to the reference class. This dramatically accelerates the reference class forecasting step in the Assess phase and reduces the systematic bias toward self-serving reference class selection.

- **Assumption surfacing:** Large language models trained on organizational documents, Decision Records, and strategic planning materials can identify implicit assumptions in new decision proposals, assumptions that are present in the logic of the proposal but have not been explicitly stated. This application of AI is particularly valuable for the "key assumptions" field of the Decision Record, where the tendency to write only the assumptions one is comfortable examining is a systematic limitation of the human process.

- **Cognitive bias flagging:** AI systems can be trained to recognize the linguistic and structural markers of specific cognitive biases in decision proposals and deliberation transcripts, the language of anchoring, the structure of confirmation bias, the framing of loss aversion, and flag them for the decision-making team's attention.

This is not a replacement for the structured debiasing protocols that Susan's framework includes; it is a complement to them, extending the reach of debiasing to decisions that do not go through the full ADEL process.

The Trolley Problem Arrives at MegaHealth

In Month 22 of the Decision Sciences program, an AI vendor's demonstration produces a moment that Jim will later describe as the most clarifying ethical experience of his executive career.

The vendor is demonstrating an AI-powered system for managing intensive care resources. During times of scarcity, this system would optimize ICU bed allocation across the entire health network in real time to maximize the expected survival rate for all patients. In simulations, the system's performance is impressive, showing statistically significant improvements in survival rates compared to traditional, manual allocation methods.

The demonstration proceeds smoothly until a clinical quality officer in the room asks a question.

Dr. Marcus Webb, VP of Clinical Quality, "In the simulation where the system optimizes survival benefit, what happens to a seventy-eight-year-old patient with heart failure and low predicted survival probability who is already admitted to the ICU when a forty-two-year-old trauma patient with high predicted survival probability arrives and needs a bed? Does the system recommend transferring the seventy-eight-year-old?"

The vendor's demonstration team pauses.

Vendor Representative, *"In the optimization model, yes. The survival benefit maximization logic would flag the seventy-eight-year-old's bed as a candidate for reallocation, and the system would recommend, not execute, a transfer discussion with the clinical team."*

There is pin drop silence in the room.

Jim, *"The system is recommending that we have a conversation with a seventy-eight-year-old patient about transferring her to a lower-acuity setting to make room for someone the algorithm has determined has a higher expected survival benefit.”*

Vendor Representative, *“The clinical team makes the final decision. The system is advisory.”*

Jim, *“The clinical team is now making that decision under the specific pressure of a system that has told them, in front of documentation that will be reviewed, that the algorithmically optimal choice is to transfer the patient. That is not equivalent to a clinical team making an independent judgment. It is a clinical team whose judgment is anchored to an algorithmic recommendation with explicit survival-benefit justification. The anchoring and the authority of the system will change how they decide, even if they retain nominal decision authority.”*

He looks around the room.

Jim, *“We are not deploying this system. Not because the survival benefit optimization is wrong. It may well produce better population-level outcomes. But the governance framework for deploying a system that explicitly recommends individual patient outcomes in ways that are sensitive to age and diagnosis, in a way that will inevitably be perceived as rationing, regardless of the clinical framing, requires a level of community engagement, ethical review, and regulatory clarity that we do not have. We will flag this for the ethics committee and revisit when we have the right governance foundation.”*

The vendor is disappointed. Marcus Webb is visibly relieved. Elena Vasquez, who is attending the demonstration as a board observer, writes two words in her notebook: “He gets it.”

CHAPTER NINE:
THE WISDOM IMPERATIVE

Where the story ends, and where the real work begins

Eighteen Months: What Changed

It is now 18 months since Susan walked into Jim's conference room with a coin and a moleskin, 12 months since the roll out, and MegaHealth is a different organization. While the transformation is not dramatic, it is a slow process measured in years rather than quarters, and anyone suggesting otherwise is likely selling something, it is meaningfully different is what Susan tells the board in this final meeting. She goes on to inform them that the changes are measurable, durable, and compound significantly over time.

The decision database contains 1.8 million logged decisions. Of these, approximately 12,000 are classified as significant, decisions involving material resource commitments, genuine alternatives, and trackable outcomes. The outcome data on the first eighteen months of significant decisions is now mature enough for Susan's team to conduct the analysis that Elena Vasquez requested at the board meeting to test whether the organization is making better decisions or merely tracking them more carefully. The finding is nuanced, honest, and genuinely encouraging.

Decisions made through the full ADEL framework, with documented assessment, formal decision record, cascade mapping, and outcome tracking, show outcome quality approximately 31 percent higher than comparable decisions made outside the framework, measured against the pre-committed success metrics specified in the original decision records. This effect is most pronounced in strategic and capital decisions, which are characterized by long feedback loops; in these areas, the benefits of systematic assessment are significant. Conversely, the impact is smallest for clinical and operational decisions, which have short feedback loops and already rely on existing protocols that provide similar cognitive forcing.

The finding is not that the framework is magic. The framework acts as a forcing function by mandating the cognitive labor necessary to produce higher-quality decisions. It ensures this work occurs at critical moments when it would otherwise be bypassed in the absence of an explicit, structural commitment.

Jim's Retrospective: What the Physician Learned

Jim has his own retrospective to conduct. He has been tracking his decisions in the moleskin, and then in the digital system, since the first Tuesday morning when Susan handed him the notebook. He has 400 personal decision records.

Looking back at them, he sees patterns he did not see at the time.

He performs better in the morning than in the afternoon, a pattern that confirms what the aggregate data shows. He is also more effective when he documents his assumptions before consulting his team than when he forms his views collaboratively. While he initially resisted this finding, he ultimately accepted it as accurate. He is better when the stakes are highest, which he finds both gratifying and illuminating: the adrenaline of genuinely high-stakes decisions activates System 2 more reliably than routine challenges.

He is consistently worse when he combines time pressure with interpersonal conflict, when a decision involves disagreement with a person he respects and must be made quickly. This is the combination that produces his most reliable cognitive failures: he defaults to the more senior person's view, or to whatever resolves the conflict fastest, rather than to the evidence.

This finding does not make Jim feel good about himself. But it does something more valuable: it allows him to design for it. High-stakes decisions that involve interpersonal conflict are now, systematically, delayed by at least 24 hours, meaning that they never make it in the same meeting where the conflict surfaces

They are always reviewed against the Decision Record framework before commitment. This simple protocol, which Jim designed for himself based on eighteen months of honest self-observation, has, by his own assessment, significantly improved the quality of about twenty of the most important decisions he has made in the past year.

He also notices something about the physician he was at thirty-five, the one who saw Gerald in a twelve-minute appointment and missed the warning signs of a myocardial infarction. The physician was not incompetent, but he was, however, operating without the cognitive infrastructure necessary to support his decision-making process. The cognitive load of the appointment environment, the time pressure, the competing demands of a full waiting room, these conditions degraded the decision process in exactly the ways that behavioral science now predicts and that Decision Sciences provides tools to counteract.

Jim is not the same decision-maker he was eighteen months ago. He is not necessarily smarter. But he is more honest about the conditions under which his judgment is unreliable, more deliberate about creating conditions that support his best thinking, and more systematic about learning from the gap between what he expected and what happened. In the language of Aristotle, he is developing *phronesis*, or practical wisdom. In the language of Susan, he is evolving into the specific type of decision-maker that the organization requires him to be.

Susan's Reflection: The Scientist Who Learned to Tell Stories

Susan has her own retrospective.

She has spent twenty years building analytical systems that were technically excellent and organizationally underutilized. She built models that were right. She built dashboards that were clear. She built recommendations that were well-supported by evidence. And she watched, with mounting frustration, as those recommendations were acknowledged and then quietly set aside in favor of whatever the room had already decided.

What changed at MegaHealth, what made this engagement different from every previous one, was not the technology. It was the recognition that data scientists who want to influence decisions must become students of the decision process itself. Not just what the data says, but how human beings hear it, interpret it, and act on it, or fail to.

She had to learn to tell stories. The coin flip, the Rock, Paper, Scissors game, the Team A and Team B experiment, and the Kelsey conversation about the Linac were not merely for entertainment. They were intentional cognitive interventions, carefully designed experiences engineered to reveal specific insights about Jim's cognitive vulnerabilities at the exact moments they were most relevant to his education.

Building them required Susan to engage as deeply with psychology, philosophy, and narrative structure as with statistics and information architecture.

This shift represents the evolution currently underway across the data science profession, a transition that many practitioners have yet to fully embrace. The era defined by pure technical excellence, where model quality was the sole arbiter of organizational outcomes, is being superseded. Today, the most effective data scientists are those who understand that producing an insight is only half the task; the true measure of success is ensuring that those insights meaningfully alter behavior. Decision Sciences serves as the essential discipline to bridge this gap.

Plato's Cave Revisited: What the Prisoners Could Have Done

We began this book in Plato's cave, with prisoners watching shadows and mistaking them for reality. It is worth returning to the allegory now, at the end of Jim and Susan's story, to ask a question that Plato's original allegory does not quite answer: what could the prisoners have done, short of escaping the cave entirely?

In The Republic, Plato argues that society requires a philosopher-king, an individual who has "seen the sun," recognizes that the shadows on the cave wall are merely illusions, and possesses the wisdom to lead the prisoners out of the cave toward genuine knowledge. This represents the archetype of an external deliverer, someone who brings enlightenment from a higher vantage point to those trapped in ignorance.

Decision Sciences offers a different answer. The prisoners could not escape the cave. They still cannot. We still cannot. The fundamental epistemic constraint, that we work with representations rather than reality, with models rather than the world, is as irreducible as ever.

What Decision Sciences provides is not escape from the cave but the systematic cultivation of epistemic humility: the habit of asking, continuously and honestly, *"How accurately does this shadow represent the object? What assumptions am I making about this shadow that I should test? What would change my mind about what I think I see?"* The Decision Record is an instrument of epistemic humility.

The pre-mortem is a practice of epistemic humility. The pre-deliberation independent estimation is epistemic humility institutionalized. Trigger-based review functions as epistemic humility institutionalized within governance. Within the ADEL framework, the learning phase serves as the organizational practice of this same humility—a structural acknowledgment that current knowledge is provisional and subject to revision as new evidence emerges.

This is what Socrates was doing in the Athenian agora, according to Plato's dialogues: not providing answers but cultivating the capacity for genuine inquiry. *"I know that I know nothing"* is not nihilism. It is the starting point for learning, the recognition that our current models of the world are provisional, that better models are available through disciplined inquiry, and that the willingness to be wrong is the precondition for being right.

Jim, eighteen months into his Decision Sciences journey, has not become omniscient. He has not eliminated the biases documented in Chapter Two. He has not built an organization that always makes the right choice. What he has built is an organization that makes choices with greater deliberateness, tracks them with greater honesty, and learns from the gap between expectation and reality with greater discipline. This is not omniscience; it is wisdom. And wisdom, as Aristotle observed, is not a destination—it is a practice.

Five Lessons for Healthcare Leaders, and Leaders Everywhere

Effective leadership in the age of AI requires moving beyond the mere accumulation of insights to fundamentally redesigning how organizations process, learn from, and act upon information. To navigate this shift, we must adopt a more deliberate approach to decision-making, as outlined in the following five essential lessons that include:

1. **The Problem Is Almost Never the Data**

    Every organization that has invested significantly in data infrastructure and remains dissatisfied with its decision quality is experiencing the same pattern: the data is not the bottleneck. The decision process is. More data will not fix a broken decision process. Better analytics will not fix a broken decision culture. The investment that most organizations need most and are least inclined to make is in the discipline of decision-making itself, the architecture, the governance, the tracking, and the learning infrastructure that converts analytical capability into actual decision improvement.

2. **Classify Before You Decide**

    The 2×2 decision matrix, Known/Unknown information crossed with Routine/Non-routine situation, is not a theoretical construct. It is a practical tool for determining the right cognitive investment for each decision. Routine, well-understood decisions should be automated or delegated to structured protocols. Novel decisions with uncertain information require the full ADEL framework, including explicit debiasing interventions. Applying the same level of deliberation to every decision is as wasteful as applying none.

3. **Build the Feedback Loop Before You Need It**

    The organizational learning loop, connecting decisions to their outcomes and feeding that connection back into future decision quality, must be built before the decisions whose outcomes you want to learn from are made. Every day without the feedback loop is a day of organizational experience that cannot be systematically learned from. The cost of building the feedback loop is the cost of the infrastructure. The cost of not building it is compounded ignorance, accumulated experience that produces the illusion of wisdom while actually generating only the confidence that comes from pattern-matching on an unexamined past.

4. **Make Decision Quality a Leadership Metric**

The organizations that advance furthest in Decision Sciences are those where the CEO and the board treat decision quality as a performance metric alongside revenue, quality, and patient satisfaction. This requires two things that most leadership teams find genuinely difficult: the willingness to evaluate the decision process separately from the decision outcome, and the willingness to be evaluated by the same standard to which they hold their organizations. Both are acts of leadership courage. Both are prerequisites for genuine Decision Sciences culture.

5. **Ethics and Governance Are Not Optional**

As AI-driven decision automation accelerates, the distance between human intention and machine action compresses. The governance and ethics infrastructure that ensures organizational decision-making serves human values, rather than the optimization objectives that algorithms are designed to maximize, becomes increasingly critical. Healthcare organizations that automate decisions without robust governance frameworks are building systems that may produce measurable efficiency gains while systematically violating the values they exist to serve. The warning label on Decision Sciences implementation: build the governance before you build the automation.

The Challenge

Healthcare operates at a unique intersection of scientific rigor and human vulnerability. Its decisions, about treatment and resource allocation, about expansion and closure, about quality and access, affect the most fundamental aspects of human life. They deserve the most rigorous decision-making architecture available.

The discipline to build that architecture, to invest in the systematic improvement of organizational judgment as deliberately as organizations invest in the improvement of their clinical processes and their financial systems, is the defining leadership challenge of the next decade. The organizations that rise to it will not merely be more efficient or more profitable. They will be more worthy of the trust that patients, communities, and employees place in them.

Jim and Susan are still at work. The Decision Dashboard grows every day. The feedback loop turns. The organization learns.

That is not a conclusion. It is an ongoing practice. And it begins, as all genuine practices begin, with the next decision.

> *"The unexamined decision, like the unexamined life, is not worth making. Not because examination guarantees good outcomes, it does not, but because examination is the only way to learn from what we choose, and learning from what we choose is the only way to become, over time, genuinely worthy of the choices we are given."*

-J. Tod Fetherling

Montaigne and the Essay as Decision Record

Michel de Montaigne, the sixteenth-century French philosopher who invented the essay as a literary form, described his project in terms that would resonate immediately with any serious Decision Sciences practitioner: *"Each man carries the complete form of the human condition within him."* Montaigne's essays were an attempt to understand the human condition by examining himself, not as an idealized type, not as a representative of any particular philosophical tradition, but as a specific, inconsistent, changing, error-prone individual who was genuinely curious about his own patterns of thought and behavior.

The essay, as Montaigne invented it, is an external record of internal deliberation, exactly what the moleskin is, and exactly what the Decision Record is. The word *"essay"* comes from the French essai, meaning *"attempt"* or *"trial."* Montaigne's essays are not definitive statements. They are honest attempts to examine a question, with full acknowledgment of uncertainty and the expectation of revision as new experience arrives.

The Decision Record is an organizational essay in Montaigne's sense. It is an attempt, not a certainty, not a plan that will survive unrevised, not a commitment to a specific outcome but a commitment to a specific process and a specific accountability. It says: this is what we believe now, on the evidence available, under the cognitive conditions we are operating in, with the acknowledgment that we may be wrong and the system that ensures we will learn if we are.

Just as Montaigne revised his essays over two decades, incorporating layers of correction and reconsideration to reveal the evolution of his thought, MegaHealth's Decision Record database functions as a twenty-year project in the same spirit. Its true value does not reside in any single entry; rather, it lies in the accumulation—the depth and texture of an organization's honest engagement with its own reasoning over time.

John Stuart Mill and the Value of Opposing Views

John Stuart Mill's *On Liberty*, published in 1859, contains what is arguably the most compelling philosophical argument for the institutional value of dissent ever written. Mill's argument is not merely that people should be free to express contrary opinions as a matter of rights, though that claim is central to his liberalism.

It is that the free expression of contrary opinions is instrumentally valuable for the quality of collective reasoning: the process of defending ideas against the strongest available objections is what makes those ideas reliable, and the process of encountering genuinely held contrary views is what prevents the intellectual complacency that Mill called the *"deep slumber of a decided opinion."*

The contrarian saint's advocate protocol in MegaHealth's decision framework is a Millian institution. Its function is not to introduce artificial disagreement for its own sake but to ensure that the dominant view is subjected to the strongest available challenge, to prevent the *"deep slumber of a decided opinion"* from producing the kind of groupthink that Janis documented in his analysis of catastrophic organizational decisions.

Mill's argument carries a vital organizational implication that his political philosophy leaves largely unexplored: the utility of contrary views depends fundamentally on the conditions under which they are expressed. In a psychologically unsafe environment, where dissent is met with punishment or social penalties, even the most sincere and valuable perspectives will be suppressed.

The psychological safety that Edmondson's research identifies as the precondition for organizational learning is, in Mill's terms, the institutional precondition for the kind of epistemic liberty that makes collective reasoning reliable. You cannot build a good decision culture without building a safe one. These are not separate projects. They are the same project at different levels of analysis.

The Zen Master's Question: What Is the Sound of One Hand Making a Good Decision?

The Zen tradition's use of the koan, the paradoxical question or statement that cannot be resolved through ordinary conceptual thinking and that pushes the practitioner toward a different quality of awareness, has an unexpected resonance with the deepest challenge of Decision Sciences: the challenge of deciding well under genuine uncertainty.

The standard approach to decision-making, and the approach that classical decision analysis prescribes, is to resolve uncertainty before deciding, gathering more information, building a better model, and improving the forecast. But there is a category of decisions, the most important, the most consequential, the ones that define an organization's strategic direction over years or decades, where the uncertainty is irreducible.

You cannot resolve it by gathering more information, because the information you need does not yet exist. You cannot model it away, because the future state of the world that your decision is supposed to produce depends on your decision and on the responses of other agents to your decision, in a way that defies clean modeling.

The Zen koan does not resolve paradoxes. It trains a quality of mind that can hold paradox without being paralyzed by it, that can act with full commitment while maintaining full awareness of uncertainty. The Japanese concept of mushin, *"no-mind,"* the state of non-attachment to outcome that allows a martial artist or a swordsman to respond with perfect timing and efficiency, is the warrior's version of the same quality.

For Decision Sciences, the practical relevance is this: the goal of the framework, the cognitive debiasing, and the process discipline is not to eliminate uncertainty. Instead, it is to cultivate a quality of organizational mind capable of holding uncertainty honestly, avoiding both the trap of false confidence and the paralysis of anxiety. It is about committing fully to the best available decision while simultaneously building the feedback loops that will reveal whether that commitment was well-placed.

Jim, at the end of his second year with the moleskin, has not resolved the uncertainty of organizational leadership. He has developed a quality of relationship to that uncertainty that is more honest, more disciplined, and more productive than the relationship he had before. He does not know more than he did. He knows better what he does not know. That, Socrates argued twenty-four centuries ago, is the beginning of wisdom. The beginning. Not the end.

The Long View: What Organizational Wisdom Actually Looks Like

The Japanese concept of shokunin, usually translated as *"artisan"* or *"craftsman"* but carrying connotations of total dedication to craft, lifelong commitment to mastery, and the subordination of personal comfort to the perfection of one's work, has no direct equivalent in Western organizational culture. The word describes something like the opposite of the modern corporate approach to developing capability: not training programs and competency frameworks, but decades of patient practice, honest self-assessment, and the gradual refinement of judgment through thousands of repetitions.

The sushi chef Jiro Ono, whose Sukiyabashi Jiro restaurant in Tokyo holds three Michelin stars and whose career was documented in the 2011 film Jiro Dreams of Sushi, trained for ten years before he was allowed to make tamago, the egg dish that is, in the Jiro tradition, the test of a chef's fundamental mastery. Not the complex preparations. The egg. Because the egg, done perfectly, requires exactly the kind of calibrated, feedback-sensitive, continuously refined judgment that distinguishes genuine mastery from technically adequate performance.

Jim, at Month 18 of his Decision Sciences journey, is not Jiro. He is not even a ten-year apprentice. But he is something important: he is a leader who has begun to practice his craft with the deliberateness that mastery requires. He is writing down his decisions. He is reviewing his predictions. He is building a feedback loop between his choices and their consequences that, over years of disciplined engagement, will produce the kind of calibrated judgment that cannot be taught but can be earned.

The organizational parallel is direct. A health system that has been running its Decision Sciences program for five years possesses an asset that a system just beginning its journey cannot purchase: a comprehensive database of thousands of decisions with tracked outcomes, a culture recalibrated toward honest self-assessment, and a set of institutional habits that carry the memory of past decisions into the future. This is organizational shokunin—the patient accumulation of practiced, reflected-upon, documented, and remembered experience that matures into genuine institutional wisdom.

What Elena Vasquez Wrote in Her Notes That Night

Elena Vasquez keeps detailed notes after every board meeting. They are private, never shared, never reviewed, but they constitute a running record of her assessments of the organizations she serves, the leaders she works with, and the ideas she finds worth remembering.

After the board meeting where Jim presented the full decision analytics report, Elena wrote three pages. The first page summarized the data. The second page assessed Jim's leadership; she had been uncertain about him when he was appointed, noting that his clinical reputation was excellent but his organizational track record was shorter than she would have liked. The assessment she wrote was the most positive she had given any CEO in twelve years of board service.

The third page was different. It was a question, the kind of question Elena asks herself when she thinks she may be looking at something genuinely new:

What are the conditions that allow an organization to become genuinely humble about its own limitations?

What makes one leadership team capable of building a system that makes it more accountable, when most leadership teams are constitutionally incapable of the genuine vulnerability that kind of accountability requires?

This is not a personality question. It is a systems question. Jim is not uniquely humble by personality. He is a man of considerable ego who has been through something that most leaders avoid: a genuine, private reckoning with the consequences of his own cognitive limitations. Gerald's case. The mistake that didn't kill the patient but forced an honest encounter with the limits of expertise. Whatever produced that reckoning in 1991, whatever it is that converts experience into wisdom rather than just confidence, is the variable that most determines whether a Decision Sciences program becomes transformation or theater.

Elena closed her notebook and thought about the question for a long time. She did not write a conclusion. She was not sure there was one. But she thought the question was important enough to keep.

The Ongoing Practice: Year Two and Beyond

The Decision Sciences program at MegaHealth does not end when this book does. It is, by design and by necessity, an ongoing practice, one that will evolve as the organization's needs evolve, as the technology infrastructure enables new possibilities, and as the cultural norms of the organization continue to be shaped by accumulated experience with the framework.

Year Two brings three significant developments that neither Jim nor Susan fully anticipated. The first is the emergence of a community of practice, a self-organizing network of Decision Sciences practitioners at the middle management level who are beginning to share techniques, compare notes, and develop local adaptations of the framework that work better for their specific contexts than the original design.

The community of practice was not planned. It emerged from the quarterly retrospective sessions, where managers from different parts of the organization began exchanging insights informally, and then formally. It is, in the language of organizational behavior, a grassroots knowledge management system, and it is producing innovations in Decision Sciences practice that Susan's team had not generated centrally.

The second development is the first genuine test of the sunset protocol. A clinical program in the Southeast Quadrant, a chronic disease management initiative that had been launched with strong evidence and full ADEL compliance, reaches its Month 18 review trigger. The outcome data is unambiguous:

- The program is producing 40 percent of its projected impact at 80 percent of its projected cost.

- The trigger condition, less than 50 percent of projected impact at Month 18, is met.

- The sunset protocol is activated.

What follows is the most important moment in the entire Decision Sciences implementation, because it is the first time the framework is tested against a real organizational emotion: the reluctance of a well-meaning, hard-working program team to acknowledge that their program has underperformed.

Jim chairs the review meeting. The program team presents their data with total transparency; concealment is effectively impossible in a system where outcome data is tracked automatically. They bring a strong recommendation to continue with modifications, arguing that the program is on a positive trajectory and that abandonment would squander the relationship capital built within the community over the past eighteen months.

Jim listens carefully. He consults the Decision Record. He asks the devil's advocate the standard questions. And then he does something that the program team does not expect.

Jim, *"I want to apply the pre-mortem to the decision to continue rather than the decision to stop. Imagine it is Month 30 and we continued the program and it still has not reached its targets. What happened?"*

The pre-mortem produces an insight that the forward-looking analysis had not surfaced: the program's underperformance is concentrated in a specific patient segment, working adults who cannot attend the in-person sessions that form the program's core. The program is working excellently for retired and part-time patients. It is not working at all for patients who work full-time.

Jim's decision: not continuation, and not termination. Redesign. The relationship capital is preserved. The underperforming component is replaced. The program is relaunched with a digital supplement for working patients. By Month 30, it exceeds its original performance targets.

The sunset protocol did not produce termination. It produced the honest assessment that termination might be warranted, which produced the level of organizational scrutiny that revealed the redesign opportunity that had been invisible while everyone assumed the program was on track. The protocol worked exactly as designed.

The Gift Susan Gives Jim on Year Two's Anniversary

On the second anniversary of that Tuesday morning meeting where she first handed him the coin, Susan sends Jim a package. Inside, he finds a new Moleskine notebook—identical in every detail to the first one, which is now completely full—and a single typed page.

The page contains a summary of Jim's decision record over two years. Not an evaluation, not a grade or a ranking. A portrait: the patterns that define his decision-making identity, assembled from two years of systematic self-observation.

It tells him that he has made 847 significant decisions, logged with the full ADEL protocol. His prediction accuracy on primary outcome metrics has improved from 58 percent in Year 1 to 67 percent in Year 2, a 16 percent improvement in predictive calibration. His confidence interval hit rate has improved from 62 to 74 percent, still below optimal but meaningfully better than where he started.

It tells him where he is still weakest: market forecasting, which continues to show systematic optimism bias; and talent management decisions, where his confidence significantly exceeds his accuracy. These are the areas where Year 3's work will concentrate.

And it tells him something that Susan has never said directly:

"Two years ago, I handed you a coin and a moleskin. I told you this was how Decision Sciences begins. What I did not tell you then, because you would not have believed it yet, is that this is also how it ends: not with certainty, not with perfection, not with a decision system that eliminates risk and guarantees outcomes. It ends with a leader who knows more precisely what he does not know, who is honest about the gap between his confidence and his accuracy, and who has built an organization that learns from the difference. You have built that. It is not finished. It will never be finished. But it is real, and it is yours."

Jim reads the page twice. He sets it on his desk, next to the note from Gerald.

He picks up the new moleskin.

He writes: Decision 1. Date: [today]. The organizational Decision Sciences program enters Year Three.

He has not yet made Year Three's first decision. He has only committed to making it deliberately, recording it honestly, and learning from whatever follows.

That, Susan has taught him, is enough.

"Wisdom is not the absence of uncertainty. It is the ability to act well in its presence."

-J. Tod Fetherling

APPENDIX A:
THE COMPLETE DECISION RECORD TEMPLATE

The Decision Record is the foundational governance artifact of the Decision Sciences practice. It is designed to capture every element of a significant decision that will be needed for future review, retrospective learning, and organizational accountability. Every field has a specific cognitive and governance purpose; none is bureaucratic overhead.

Field	Purpose and Instructions
Decision ID	System-generated unique identifier. Format: [ORG]-[YEAR]-[SEQUENCE] (e.g., MH-2025-0047). Enables cross-reference and longitudinal tracking.
Decision Title	Clear, action-oriented description in 10 words or fewer. Should describe what was committed, not what was considered.
Decision Owner	Single individual who is ultimately accountable. No shared ownership, one owner per decision, no exceptions.
Decision Date	The date the commitment was formally made. Not the date analysis began or the date implementation started.

Decision Type	Choose from: Strategic / Capital / Clinical / Operational / Technology / People / Financial / Regulatory.
Urgency Level	Time-Sensitive (must decide within 48 hours) / Plannable (1-4 weeks) / Deferrable (can wait without cost).
Decision Statement	Precisely what was decided, in 2-3 sentences. Specific enough that a reader with no prior context understands exactly what commitment was made.
Alternatives Considered	All options that received genuine analysis, not just the chosen option. For each alternative, note why it was not selected.
Selection Rationale	The specific evidence, analysis, and judgment that supports the chosen option over alternatives. Should be falsifiable, it should be possible to specify what evidence would have changed the decision.
Key Assumptions	What must be true for this decision to produce the expected outcome? List 3-7 specific, testable assumptions. These will be the focus of the outcome review.
Success Metrics	How will outcome quality be measured? Specify quantitative metrics, measurement methodology, and timeline for each.
Review Triggers	Conditions that should prompt reassessment, regardless of scheduled review dates. What would have to change in the environment for this decision to require reconsideration?
Stakeholders Consulted	Who provided meaningful input to this decision. Distinct from the RACI Accountable, this documents intellectual contribution, not organizational accountability.
Biases Checked	Which specific cognitive biases were explicitly addressed in the decision process? (Anchoring, confirmation, availability, planning fallacy, etc.)

Confidence Level	Decision-maker's confidence in the selection rationale, on a scale of 1-10, with brief explanation. This pre-commitment enables calibration analysis at outcome review.
Outcome Record (completed at review)	What actually happened, measured against the pre-committed success metrics. Honest assessment of variance and its causes.
Learning Record (completed at review)	What this outcome reveals about the accuracy of the key assumptions, the validity of the selection rationale, and the reliability of the decision process used.

EEBRI: A Personality Framework for Decision-Making

The EEBRI concept (Effective and Efficient Behavior by Rational Individuals), derived from Modelnetics, provides a personality-based lens for analyzing individual decision-making styles within an organization. Because different cognitive and behavioral profiles produce systematically different decision tendencies, mapping the dominant profile of a leadership team allows us to design decision processes that specifically compensate for their predictable weaknesses.

Profile	Decision Style	Strength	Primary Bias Risk	Design Response
Analytical	Data-intensive; slow; seeks certainty before committing	Rigor and accuracy	Analysis paralysis; anchoring on data	Add decision forcing functions and time constraints
Driver	Fast, confident; trusts System 1; low tolerance for ambiguity	Speed and decisiveness	Overconfidence; insufficient analysis	Build mandatory pre-mortem and assumption surfacing into protocol
Expressive	Collaborative; consensus-seeking; social signals heavily weighted	Alignment and buy-in	Groupthink; social proof; authority bias	Require independent pre-deliberation estimation; rotate devil's advocate
Amiable	Conflict-avoidant; defers to authority; risk-averse	Relationship preservation	Status quo bias; sunk cost; loss aversion amplified	Explicit loss framing in analysis; separate evaluation from social context

APPENDIX B: THE COFFEE MAKER CASE STUDY

Across-industry illustration of Decision Sciences methodology applied to a product portfolio investment decision, the example Susan uses to build Jim's decision muscle before applying the framework to healthcare stakes.

Susan to Jim, *"Let me make you a Product Manager for a consumer goods company for twenty minutes. You sell coffee makers. You have an $8 million development budget and two products competing for it: a single-serve platform and a drip coffee maker platform. How do you decide?"*

Jim, *"Well, which one is more profitable?"*

Susan, *"That's an output, not an input. Let's build the analysis. Same framework, different industry. Same cognitive challenges, different content."*

The Pragmatic Marketing Framework Applied to Coffee

The Pragmatic Marketing Framework, a product management methodology widely used in technology and consumer products, structures investment decisions across three domains: Market (what customers need and what the competitive landscape looks like), Strategy (how the product competes and positions), and Execution (what it takes to deliver the product successfully). Applied to the coffee maker decision:

Domain	Single-Serve	Drip Coffee
Market Size	$4.2B growing at 12% annually	$2.8B declining at 2% annually
Buyer Profile	Individual household, office, convenience-seeker	Family household, value-seeker, coffee enthusiast
Competitive Position	Premium; strong brand differentiation possible	Commodity pressure; difficult differentiation
Strategy: Positioning	Convenience + personalization; premium pricing	Value + reliability; price competition
Execution: Complexity	Pod ecosystem; supply chain complexity	Simple product; well-understood manufacturing
5-Year NPV (Base Case)	$12.4M	$4.1M
5-Year NPV (Conservative)	$7.2M	$1.8M

The Pre-Mortem for Each Option

Single-Serve failure modes: Keurig dominates the market so completely that differentiation is impossible. Pod ecosystem costs spiral beyond projections. Sustainability concerns from single-use pods create regulatory or reputational risk. Retailer consolidation reduces shelf access.

Drip failure modes: Commodity pricing pressure eliminates margin despite development investment. Target market demographic (families) shifts to single serve as household income rises. Chinese manufacturing competition undercuts on price. The innovation cycle is too slow to justify development costs.

Susan, *"Notice what the pre-mortem reveals. The single-serve failure modes are mostly competitive and market risks, things that could happen but are not inevitable. The drip failure modes are mostly structural, the market is declining, the margin is eroding, and these are trends, not risks. The pre-mortem makes visible that the decision is not really between two products. It is between a competitive challenge and a structural decline. That is a much easier decision."*

Decision	Invest $6M in single-serve platform acceleration; maintain $2M in drip platform stability investment.
Rationale	Single-serve market trajectory, margin profile, and differentiation potential all favor investment. Drip maintenance investment protects existing revenue during transition without committing to a declining market.
Key Assumption	Single-serve premium positioning is achievable given Keurig's dominance. Test within 12 months via retail buyer meetings and consumer research.
Success Metric	20% improvement in single-serve platform NPS within 18 months; hold drip revenue flat year-over-year.
Review Trigger	If single-serve new product NPS does not reach 7.5 within 12 months of launch, revisit platform strategy.

Susan, *"Jim, the coffee maker decision is intellectually identical to the Linac decision. Different industry. Different stakes. Same cognitive challenges: anchoring on the status quo, availability bias from what you recently experienced in the market, overconfidence in the base-case projections, confirmation bias toward whichever option your team is already leaning toward. The framework strips all of that away and gives you the decision that the evidence supports. That is all Decision Sciences does. It removes the noise so you can hear the signal."*

APPENDIX C: SELECTED BIBLIOGRAPHY AND FURTHER READING

Foundational Philosophy

- Plato. The Republic. Translated by G.M.A. Grube, revised by C.D.C. Reeve. Hackett Publishing, 1992.

- Plato. Meno; Phaedo; Phaedrus. Multiple translations. Plato's treatments of knowledge, justified belief, and the nature of good judgment remain among the most penetrating analyses in the philosophical tradition.

- Aristotle. Nicomachean Ethics. Translated by Terence Irwin. Hackett Publishing, 1999. Essential for understanding phronesis (practical wisdom) as distinct from theoretical knowledge and technical skill.

- Marcus Aurelius. Meditations. Translated by Gregory Hays. Modern Library, 2002. The most intimate surviving record of a leader's decision-making practice.

- Epictetus. Discourses and the Enchiridion. Multiple translations. The Stoic framework for distinguishing what is within our control, critical for managing outcome uncertainty.

- Descartes, René. Discourse on the Method. 1637. The foundational text of the rationalist tradition whose limitations Damasio and behavioral economics have documented.

- Kant, Immanuel. Groundwork for the Metaphysics of Morals. Translated by Mary Gregor. Cambridge University Press, 1997.

- Nietzsche, Friedrich. Thus Spoke Zarathustra. Translated by Walter Kaufmann. Penguin, 1978.

- James, William. Principles of Psychology. 1890. Dover Publications, 1950. James's treatment of habit remains among the most practically applicable in the psychological literature.

- Dewey, John. How We Think. D.C. Heath, 1910. The learning cycle that underlies the ADEL framework.

Behavioral Economics and Decision Psychology

- Kahneman, Daniel. Thinking, Fast and Slow. Farrar, Straus and Giroux, 2011. The most accessible and comprehensive treatment of dual-process theory and behavioral economics.

- Kahneman, Daniel, and Amos Tversky. "Judgment Under Uncertainty: Heuristics and Biases." Science, 185(4157), 1124-1131 (1974). The foundational paper of behavioral decision research.

- Kahneman, Daniel, and Amos Tversky. "Prospect Theory: An Analysis of Decision Under Risk." Econometrica, 47(2), 263-292 (1979). Nobel Prize-winning work on loss aversion and probability weighting.

- Thaler, Richard, and Cass Sunstein. Nudge: Improving Decisions About Health, Wealth, and Happiness. Yale University Press, 2008.

- Ariely, Dan. Predictably Irrational: The Hidden Forces That Shape Our Decisions. HarperCollins, 2008.

- Damasio, Antonio. Descartes' Error: Emotion, Reason, and the Human Brain. Putnam, 1994. Essential reading for understanding the somatic marker hypothesis.

- Baumeister, Roy, and John Tierney. Willpower: Rediscovering the Greatest Human Strength. Penguin Press, 2011. Decision fatigue and ego depletion.

- Gigerenzen, Gerd. Gut Feelings: The Intelligence of the Unconscious. Viking, 2007. A valuable counterweight to the bias-and-heuristics tradition, arguing for the adaptive value of fast-and-frugal heuristics.

- Mlodinow, Leonard. The Drunkard's Walk: How Randomness Rules Our Lives. Vintage, 2009.

- Silver, Nate. The Signal and the Noise: Why So Many Predictions Fail, But Some Don't. Penguin Press, 2012.

Organizational Behavior and Leadership

- Edmondson, Amy. The Fearless Organization: Creating Psychological Safety in the Workplace for Learning, Innovation, and Growth. Wiley, 2018.

- Janis, Irving. Victims of Groupthink. Houghton Mifflin, 1972. The definitive analysis of collective decision failure in cohesive groups.

- Simon, Herbert. Administrative Behavior: A Study of Decision-Making Processes in Administrative Organization. 4th ed. Free Press, 1997.

- Hackman, J. Richard. Leading Teams: Setting the Stage for Great Performances. Harvard Business School Press, 2002.

- Dweck, Carol. Mindset: The New Psychology of Success. Random House, 2006.

- Senge, Peter. The Fifth Discipline: The Art and Practice of the Learning Organization. Doubleday, 1990.

- Collins, James. Good to Great: Why Some Companies Make the Leap... and Others Don't. HarperBusiness, 2001.

- Sunstein, Cass, and Reid Hastie. Wiser: Getting Beyond Groupthink to Make Groups Smarter. Harvard Business Review Press, 2015.

Healthcare Decision Sciences

- Gawande, Atul. The Checklist Manifesto: How to Get Things Right. Metropolitan Books, 2009.

- Groopman, Jerome. How Doctors Think. Houghton Mifflin, 2007.

- Wennberg, John. Tracking Medicine: A Researcher's Quest to Understand Health Care. Oxford University Press, 2010.

- Graber, Mark L., Robert M. Wachter, and Christine K. Cassel. "Bringing Diagnosis Into the Quality and Safety Equations." JAMA, 308(12), 1211-1212 (2012).

- McGlynn, Elizabeth A., et al. "The Quality of Health Care Delivered to Adults in the United States." New England Journal of Medicine, 348(26), 2635-2645 (2003).

- Danziger, Shai, et al. "Extraneous Factors in Judicial Decisions." Proceedings of the National Academy of Sciences, 108(17), 6889-6892 (2011). The judicial decision fatigue study.

Decision Analysis and Strategy

- Raiffa, Howard. The Art and Science of Negotiation. Harvard University Press, 1982.

- Hammond, John, Ralph Keeney, and Howard Raiffa. Smart Choices: A Practical Guide to Making Better Decisions. Harvard Business Review Press, 1999.

- Klein, Gary. Sources of Power: How People Make Decisions. MIT Press, 1998. The research behind the pre-mortem technique.

- Flyvbjerg, Bent. "Curbing Optimism Bias and Strategic Misrepresentation in Planning: Reference Class Forecasting in Practice." European Planning Studies, 16(1), 3-21 (2008).

- Russo, J. Edward, and Paul J.H. Schoemaker. Winning Decisions: Getting It Right the First Time. Doubleday, 2002.

- Heath, Chip, and Dan Heath. Decisive: How to Make Better Choices in Life and Work. Crown Business, 2013.

- Sunstein, Cass. How Change Happens. MIT Press, 2019.

APPENDIX D:
THE PRACTITIONER'S FIELD GUIDE

Tools, exercises, and protocols for building a Decision Sciences culture in your organization

The following tools are drawn from MegaHealth's Decision Sciences implementation and from the broader research literature. They are presented in the order in which Susan introduces them to Jim, from the individual cognitive tools that any leader can deploy alone, to the organizational infrastructure that requires institutional commitment.

Individual Tools: Building Your Own Decision Discipline

Tool 1: The Personal Decision Record, Getting Started

Begin with the minimum viable practice: write down your significant decisions. The definition of *"significant"* can start broad, any decision that will consume more than an hour of your time to implement, or that affects more than two people, or that involves a genuine choice among alternatives. Write each one with:

- The decision statement (what you committed to)

- The date

- The alternatives you considered but didn't choose

- The primary reason for your choice

- What you expect to happen, and when you'll know if you were right

That is it. Five elements, ten minutes per decision. The purpose in the first thirty days is not analysis; it is the habit. The value of the practice is not visible until you review thirty days of entries and see the patterns in your own decision-making that you could not have seen without the external record.

Tool 2: The Confidence Calibration Exercise

Once a quarter, identify twenty facts that are relevant to your professional domain and that are verifiable. For each, write down your answer and a 90 percent confidence interval (the range within which you are 90 percent certain the true answer falls). Then look up the answers. Calculate what percentage of your intervals captured the true answer.

A perfectly calibrated person captures the true answer 90 percent of the time. Most professionals capture it 50–70 percent of the time. Your score tells you how much to discount your stated confidence across different domain areas, and it improves with practice.

Tool 3: The Pre-Mortem, Thirty Minutes That Save Millions

For any significant decision, before finalizing your commitment, conduct a pre-mortem in the following format:

1. State the decision as though it has been made.

2. Say: "Imagine it is eighteen months from now and this decision has failed spectacularly. What happened?" Allow three to five minutes of individual written brainstorming.

3. Collect the failure scenarios (in groups, share without attribution first to avoid anchoring).

4. For each significant failure scenario, ask: "What assumption embedded in our current plan would have to be wrong for this to happen?"

5. Prioritize the assumptions by their importance to the decision and their current level of verification.

6. Assign actions to verify the most important unverified assumptions before finalizing the commitment.

Tool 4: Reference Class Forecasting, The Outside View in Practice

For any major investment or initiative with a projected timeline and budget, apply reference class forecasting before accepting the internal projection:

1. Identify the reference class: What category of project is this? (Technology implementation; market expansion; facility construction; clinical program launch; etc.)

2. Find the data: What does the historical record show for actual outcomes of projects in this reference class? (Industry associations, academic literature, internal history if available.)

3. Apply the correction: If the median project in this reference class exceeds its budget by 28 percent, your project's base case should include a 28 percent contingency, not a 10 percent one.

4. Identify what makes your project different from the reference class, in both directions. What specifically makes you believe your project will outperform the base rate, and what evidence supports that belief?

Organizational Tools: Building the Infrastructure

Tool 5: The Decision Architecture Matrix

Before building a full Decision Sciences infrastructure, classify your organization's decision types using the following two-dimensional framework:

Decision Type	Reversibility	Complexity	Recommended Protocol
Routine Operational	High (easily reversed)	Low (well-understood)	Standardize to protocol; delegate; automate where appropriate
Structured Strategic	Low (significant sunk cost)	Moderate (familiar domain)	Abbreviated ADEL: pre-mortem + Decision Record + basic cascade
Complex Strategic	Low (significant commitment)	High (novel domain)	Full ADEL with structured deliberation and independent estimation
Emergency	Varies	High (time-critical)	Recognition-Primed Decision with post-hoc documentation and review

Tool 6: The Structured Deliberation Protocol

For complex strategic decisions, use the following deliberation protocol:

1. [Before the meeting] Each participant independently completes a Decision Assessment Form: their individual estimate of the primary outcome metric, the three most important risks, and their confidence level.

2. [Opening] Collect and aggregate the individual assessments before any group discussion. Display the distribution of estimates (not identified by participant) to the group.

3. [Contrarian Saint/Devil's Advocacy] One assigned participant presents the strongest possible case against the emerging consensus option. This is a role, not an opinion.

4. [Deliberation] Open discussion, with the facilitator responsible for ensuring that all significant uncertainties and assumptions identified in the individual assessments receive discussion time.

5. [Assumption Check] Before finalizing: "If this assumption turns out to be wrong, does that change our decision?" Work through the three to five most important assumptions.

6. [Decision and Record] The decision-maker states the decision and the primary rationale. The facilitator completes the Decision Record in the meeting, not after.

Tool 7: Building Psychological Safety for Decision Quality

The technical tools in this guide are only as effective as the cultural environment in which they are used. The single most important cultural investment a leader can make is in psychological safety, the belief that it is safe to speak up, dissent, and report bad news without social penalty.

The specific behaviors that research (Amy Edmondson's work at Harvard Business School) shows to be most effective in building psychological safety:

- Be explicitly curious rather than critical when hearing bad news: *"Tell me more about what happened"* rather than *"How did this occur?"*

- Acknowledge your own uncertainty and limitations publicly, particularly in domains where your authority makes others reluctant to challenge you

- Respond visibly and positively the first time someone challenges your stated position with evidence, this single moment can define the psychological safety climate for months

- Create structural mechanisms for dissent (anonymous input tools, rotating devil's advocate roles, explicit pre-mortem protocols) that do not require individuals to bear personal social risk to surface contrary information

- Model the review of your own past predictions honestly, including the ones that were wrong

Tool 8: The Learning Review Protocol

The learning loop, the connection between decisions and their outcomes that enables organizational improvement, is the element most often omitted from Decision Sciences implementations. The following protocol ensures that learning reviews produce genuine insight rather than retrospective rationalization:

1. [Before the review] Each participant individually reviews the original Decision Record and writes their assessment of: (a) what happened, (b) why the outcome differed from the prediction, and (c) what the decision process should do differently next time.

2. [The comparison] Present the actual outcomes against the pre-committed success metrics. Focus first on the factual comparison, before any interpretation.

3. [The counterfactual] Ask explicitly: "What would have happened if we had made the alternative decision?" This question prevents the learning from being dominated by outcome bias.

4. [The assumption review] Review each key assumption from the Decision Record. Which ones were correct? Which were wrong? What would have had to change in the assessment process for the wrong ones to have been identified earlier?

5. [The process review] Assess the quality of the decision process independently from the outcome quality. Even if the outcome was good, were there process gaps that produced lucky rather than skilled results?

6. [The learning record] Document the specific changes to the decision process that this review suggests and build them into the next decision of this type.

Starting Tomorrow: A Practical Sequence

The full MegaHealth Decision Sciences implementation, with its 1.8 million decision records, its board-level governance restructuring, and its AI governance taxonomy, took two years to build. Most organizations will not build it in that time frame or at that scale.

But you can start tomorrow with three things:

1. [Tomorrow morning] Write down one important decision you need to make in the next thirty days. Apply the pre-mortem to it, thirty minutes, alone. Write down what you learn.

2. [This week] Find one decision you made in the past six months whose outcome you can now assess. Compare what actually happened to what you expected. Write down what the comparison reveals about your prediction process.

3. [This month] Hold one structured deliberation session using the protocol in Tool 6. Any significant decision will do. Notice what the structure reveals that unstructured discussion would have missed.

These three actions will not build a Decision Sciences culture. But they will produce evidence, specific, personal, inarguable evidence, of whether the tools in this book work for you, in your context, with your specific decision challenges. That evidence is the only foundation on which genuine organizational commitment can be built.

Jim started with a coin and a moleskin.

So can you.

AUTHOR'S NOTE

This book began with a question that a CEO asked me after a consulting engagement in 2018: "We have better data than we've ever had. Why do I still feel like I'm guessing?"

The question is not unusual. I have heard versions of it in every organization I have worked with over the past two decades. The gap between analytical capability and decision quality is one of the most consistent and most consequential problems in modern organizational life, and it is one that the analytics industry, for understandable commercial reasons, has been slow to name directly. Selling better data is easier than arguing that the problem is not the data.

The book that resulted from that CEO's question, and from the research, conversations, and implementations that followed, is not primarily a book about healthcare, though it uses healthcare as its primary setting. It is a book about the discipline that connects what organizations know to what organizations do. Decision Sciences is that discipline. It is genuinely multidisciplinary, genuinely practical, and genuinely urgent. I have tried to write it in a way that honors all three of those qualities.

A Note on the Research

The behavioral economics and decision science research summarized in this book is genuine. Kahneman and Tversky's work, Thaler and Sunstein's nudge framework, Klein's naturalistic decision-making theory, Tetlock's superforecaster research, Edmondson's psychological safety work—each of these is a legitimate, peer-reviewed, empirically supported body of research. I have summarized these bodies of research as accurately as the narrative demands, and I have included citations in Appendix C for readers interested in engaging with the primary sources.

Where I have simplified for accessibility, I have aimed to preserve the core insight rather than distort it. Behavioral economics is a complex and debated field; it faces its own replication crisis, debates over external validity, and ongoing refinements. The summary here is a starting point, not a final word.

My framework, Assess → Decide → Execute → Learn (ADEL) model, emphasizes decision charters, RAPID governance, cascade decisions, and structured deliverables at each phase.

This Decision Sciences Framework (DSF) is significantly different from these popular decision frameworks in scope, depth, and purpose. Here is how they compare: The chart below illustrates the main dynamic: popular frameworks tend to cluster in the narrow and shallow corner, while the DSF spans the full lifecycle with deeper analysis. Here's a detailed breakdown:

What popular frameworks aim to do

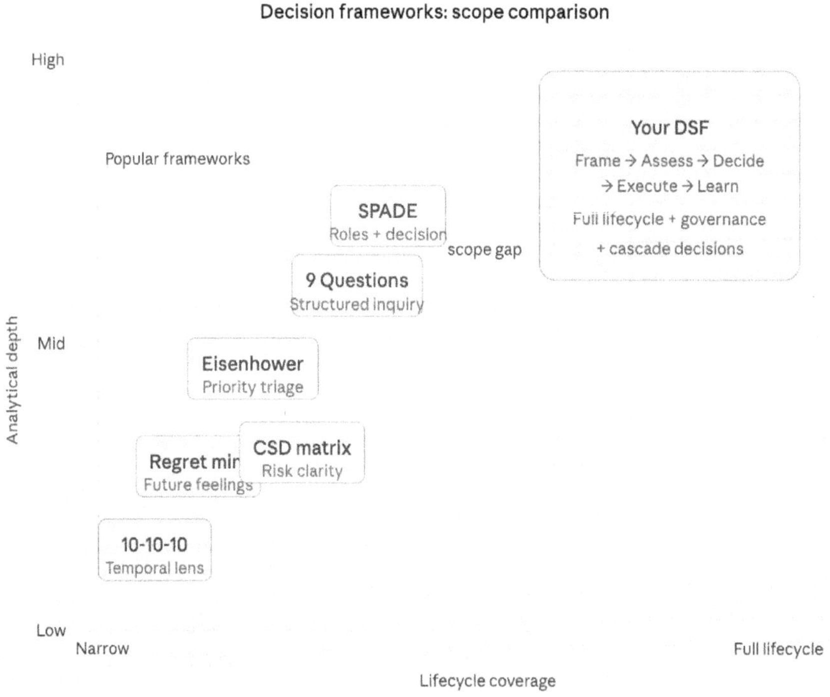

Each of the six frameworks addresses a specific, limited problem at the time of decision. The Eisenhower Matrix helps prioritize tasks. The 10-10-10 Rule counters short-term thinking by encouraging temporal perspective. Regret Minimization offers an emotional anchor when choices seem equally appealing. CSD Matrix highlights known versus assumed information. SPADE, the most structured among them, clarifies roles and reduces consensus bottlenecks. The 9 Questions Framework is essentially a preflight checklist to ensure key considerations are addressed before committing.

These tools are primarily point-in-time aids. They help move from "I need to decide" to "I've decided" more thoroughly. They don't guide what comes before or after that decision.

How the Decision Sciences Framework differs

Your DSF treats decision-making as a lifecycle, not an event. The four phases of the ADEL model (Assess, Decide, Execute, Learn) embody a different assumption: that decision quality depends not just on the choice made but also on how well the problem was defined, evidence gathered, accountability established, and lessons learned.

Several structural elements in the DSF have no equivalent in popular frameworks:

The Decision Charter (Frame phase) is the DSF's most distinctive pre-decision element. It mandates explicit agreement on what decision is being made, who has authority (RAPID model), and which constraints are non-negotiable before analysis begins. None of the six popular frameworks do this. SPADE's "People" and "Decide" components come close but do not formalize a charter document.

The cascade decisions concept (Execute phase) is entirely absent from popular frameworks. It recognizes that major strategic decisions trigger numerous supporting decisions that also must be governed, since most failures occur downstream rather than at the primary decision. For example, the MegaHealth cardiology expansion involved decisions about rent versus owning, hiring, contracting, branding, and roughly 40 other consequential choices, each requiring discipline.

The structured deliverables at each phase (Options Analysis, Decision Record, Execution Roadmap, Retrospective) build organizational memory. Popular frameworks yield a decision; the DSF produces documentation. This distinction matters, especially when future accountability is involved.

The multi-tool analytical approach in the Assess phase—covering demographic analysis, competitive positioning, financial scenarios, stakeholder mapping, and weighted scoring—far exceeds the scope of any single popular framework. These tools can be integrated within the Assess phase of the DSF; they are not mutually exclusive.

Where the popular frameworks retain value

They excel in speed and ease of use. The 10-10-10 Rule can reframe a decision in five minutes. The Eisenhower Matrix can reorganize a leader's week in a single meeting. Regret Minimization works well for high-stakes personal or entrepreneurial decisions with limited data. SPADE is particularly helpful in startup environments with informal governance and risk of paralysis from committee decision-making.

The DSF requires organizational infrastructure—a decision sponsor, facilitator, time, stakeholder engagement—which makes it less suitable for quick or low-stakes decisions. Popular frameworks serve those needs better.

An honest comparison

The DSF is to popular decision frameworks what a clinical trial protocol is to a doctor's intuition. Intuition is faster, more portable, and often correct. Protocols are slower, resource-heavy, but produce evidence that withstands scrutiny, transfers to others, and accumulates value. Both are essential; they operate at different levels of rigor and organizational stakes.

The popular frameworks can be incorporated into the DSF (mainly in Assess and Decide). I don't believe the DSF can be embedded into them because it is fundamentally larger.

A Note on Jim and Susan

Jim and Susan are composites. Every person I have worked with over twenty years of healthcare consulting and technology leadership has contributed something to one or both, a question asked, a decision made, a failure survived, an insight generated. They are not real people, but they are made of real people.

The relationship between them, the physician-executive and the data scientist, the institutional authority and the analytical outsider, the System 1 and the System 2, is the relationship that I have found, in various forms, at the heart of every successful organizational Decision Sciences program. It is not a simple relationship. It is not always comfortable. But it is productive in a way that neither party could be independent.

If you recognize yourself in either of them, if Jim's gut-feel-over-evidence habits feel familiar, or if Susan's frustration at watching good analysis evaporate into bad decisions resonates, then the book has done part of its work.

A Note on What Comes Next

Decision Sciences is a relatively young field considering the size and urgency of the issues it tackles. The behavioral economics research that underpins its scientific basis is only about sixty years old. Its organizational applications are even younger. The artificial intelligence breakthroughs that are transforming organizational decision-making have emerged more quickly than the governance frameworks needed to manage them responsibly.

There is a significant amount of work ahead— in research, practical application, governance, and the cultural shift necessary to make decision quality a true organizational priority rather than just a rhetorical goal. This book invites you to join that effort: to apply honest self-assessment and systematic learning to the decisions that most impact the lives of the people you serve.

Jim started with a coin and a moleskin. Start wherever you can.